University Texts in the Mathematical Sciences

Editors-in-Chief

Raju K. George, Department of Mathematics, Indian Institute of Space Science and Technology, Valiamala, Kerala, India

S. Kesavan, Department of Mathematics, Institute of Mathematical Sciences, Chennai, Tamil Nadu, India

Sujatha Ramdorai, Department of Mathematics, University of British Columbia, Vancouver, BC, Canada

Shalabh, Department of Mathematics and Statistics, Indian Institute of Technology Kanpur, Kanpur, Uttar Pradesh, India

Associate Editors

Kapil Hari Paranjape, Department of Mathematics, Indian Institute of Science Education and Research Mohali, Mohali, Chandigarh, India

K. N. Raghavan, Department of Mathematics, Institute of Mathematical Sciences, Chennai, Tamil Nadu, India

V. Ravichandran, Department of Mathematics, National Institute of Technology Tiruchirappalli, Tiruchirappalli, India

Riddhi Shah, School of Physical Sciences, Jawaharlal Nehru University, New Delhi, Delhi, India

Kaneenika Sinha, Department of Mathematics, Indian Institute of Science Education and Research, Pune, Maharashtra, India

Kaushal Verma, Department of Mathematics, Indian Institute of Science Bangalore, Bengaluru, Karnataka, India

Enrique Zuazua, Department of Mathematics, Friedrich-Alexander-Universität Erlangen-Nürnberg (FAU), Erlangen, Germany

Textbooks in this series cover a wide variety of courses in mathematics, statistics and computational methods. Ranging across undergraduate and graduate levels, books may focus on theoretical or applied aspects. All texts include frequent examples and exercises of varying complexity. Illustrations, projects, historical remarks, program code and real-world examples may offer additional opportunities for engagement. Texts may be used as a primary or supplemental resource for coursework and are often suitable for independent study.

Shashi Mohan Srivastava

An Introduction to Naïve Set Theory and Its Applications

 Springer

Shashi Mohan Srivastava
Indian Statistical Institute
Kolkata, West Bengal, India

ISSN 2731-9318 ISSN 2731-9326 (electronic)
University Texts in the Mathematical Sciences
ISBN 978-981-97-4642-2 ISBN 978-981-97-4643-9 (eBook)
https://doi.org/10.1007/978-981-97-4643-9

Mathematics Subject Classification: 26Axx, 30Bxx, 54Axx, 12Exx, 28Axx

© The Editor(s) (if applicable) and The Author(s), under exclusive license to Springer Nature Singapore Pte Ltd. 2024

This work is subject to copyright. All rights are solely and exclusively licensed by the Publisher, whether the whole or part of the material is concerned, specifically the rights of translation, reprinting, reuse of illustrations, recitation, broadcasting, reproduction on microfilms or in any other physical way, and transmission or information storage and retrieval, electronic adaptation, computer software, or by similar or dissimilar methodology now known or hereafter developed.
The use of general descriptive names, registered names, trademarks, service marks, etc. in this publication does not imply, even in the absence of a specific statement, that such names are exempt from the relevant protective laws and regulations and therefore free for general use.
The publisher, the authors and the editors are safe to assume that the advice and information in this book are believed to be true and accurate at the date of publication. Neither the publisher nor the authors or the editors give a warranty, expressed or implied, with respect to the material contained herein or for any errors or omissions that may have been made. The publisher remains neutral with regard to jurisdictional claims in published maps and institutional affiliations.

This Springer imprint is published by the registered company Springer Nature Singapore Pte Ltd.
The registered company address is: 152 Beach Road, #21-01/04 Gateway East, Singapore 189721, Singapore

If disposing of this product, please recycle the paper.

To my children
Suraj, Rosy, Ravi and Deepali

Preface

In order to solve a problem, "Can a function have more than one representation by a trigonometric series?" the great Russian-born German mathematician Georg Cantor made a very basic contribution to mathematics. He extended the fundamental method of induction well beyond natural numbers. In order to make his method precise, he had to consider the abstract notion of sets as an arbitrary collection of objects and functions as an arbitrary assignment of a point in a set to a point in another set. In order to state his results, he defined a real number as an equivalence class of Cauchy sequences of rational numbers and developed the topology of the set of real numbers.

Before Cantor arrived on the scene, progress in mathematics was often being achieved at the expense of rigor. For instance, the only treatment of real numbers was the geometric one as points on a line and the only functions that were being considered were those which had analytic expressions. Cantor's work brought rigor and precision in mathematics.

To study the nature of the sets of real numbers that he discovered Cantor developed such concepts as ordinal and cardinal numbers. Before Cantor's discoveries, a set was either finite or infinite. Cantor introduced the notion of number of elements in an infinite set and showed that not all infinite sets are of the same size. The reader is referred to [17] for a historical account of Cantor's discovery of set theory and topology.

Cantor's revolutionary discoveries met with serious criticisms from some of the leading mathematicians of his time. However, some great mathematicians, such as Hilbert and Dedekind, could see the usefulness of Cantor's work. Since Cantor's definition of sets was somewhat informal, several paradoxes such as Russell's paradox, Burali-Forti paradox, etc. also came to the surface. This led to the laying down of the foundations of mathematics. Today set theory is the lingua franca of mathematics. It has led to fruitful generalizations and opened up new possibilities. The great German mathematician David Hilbert was prophetic when he declared, *"No one shall expel us from the paradise that Cantor has created."*

About This Book

The aim of this book is to introduce the basic concepts and techniques of naive set theory, which are often used in other branches of mathematics. This is essentially introductory in nature and is meant as a stepping stone to modern set theory for those who would like to work in it. Also, this is the minimum amount of set theory that all mathematicians ought to know.

The book can be divided in three parts. The first part consists of Chaps. 1, 6 and 7. In Chap. 6 we briefly give some background in mathematical logic. Zermelo-Frankel axioms for set theory including the axiom of choice are presented in Chap. 7. These are used basically to give a broad idea of how the set theory developed in the book can be made precise. Chapter 1 should be read in conjunction with Chaps. 6 and 7 to develop an idea of how set theory is developed precisely.

Chapters 2 and 3 constitute the second part of the book where we introduce naive set theory to the readers. Not much prerequisites are required in this part. Some minimal knowledge of analysis and a little knowledge of general topology will be sufficient. The first two parts should serve as a stepping stone for those who want to specialize in set theory.

Chapters 4 and 5 constitute the third part of the book. In Chap. 4, we give applications of set theory that we have introduced to several branches of mathematics including real and complex analysis, linear algebra and algebra, topology, functional analysis, and measure theory. A basic familiarity with these subjects is assumed here. Chapter 5 is entirely devoted to the famous Banach-Tarski paradox.

In Chap. 2, we introduce the axiom of choice and Zorn's lemma and use them to prove further results in set theory. We also study well-ordered sets and transfinite induction on well-ordered sets in Chap. 2. Here, we give a motivated history of why and how Cantor was led to discover set theory, mathematically define real numbers and develop its topology. The reader may see [17] for this very fascinating story.

Chapter 3 introduces ordinal and cardinal numbers. Transfinite induction on ordinal numbers and cardinal arithmetic are two fundamental techniques introduced in this chapter.

The reader will find a more transparent and probably new proof of Hahn decomposition theorem on signed measures in the section on measure theory in Chap. 4. Chapter 5 is devoted to the proof of Haudorff's theorem, *"there does not exist a finitely additive measure on the power set of $\mathbb{R}^n$, $n \geq 3$, such that the measure of the unit cube is one and whenever two sets A and B are congruent, there measures are equal."* The proof exploits beautifully the richness of the group of rotations around lines passing through the origin when $n \geq 3$. Hausdorff's proof leads to the famous Banach-Tarski Paradox—*"the closed unit disc of $\mathbb{R}^3$ can be partitioned into finitely many sets and then they can be put together in such a way as to get two closed discs of the same size as the closed unit disc."* This is also presented in full detail in Chap. 6.

How to Use the Book

This is primarily an introductory textbook in naive set theory for senior undergraduate students. Depending on the background, an instruction may first give a quick working introduction to mathematical logic from Chap. 6 followed by presenting the axioms of ZF from Chap. 7. Chapters 1–3 constitute the main part of the book. It is meant as an introductory course in naive set theory for senior undergraduate students. Chapter 5 on Banach-Tarski paradox is optional. A little background in mathematics is needed for this.

Chapter 4 is devoted to the applications of set theory. Most of it requires serious background from mathematics. It is helpful to develop grip on set theory. It is not expected that a student will have the entire background from mathematics. Depending on the background of the students, an instructor may pick some topics to teach them. Chapter 4 can also be used as reference by mathematicians from different background. A large number of applications from several areas of mathematics is presented.

Exercises are integral part of the book. Students must workout problems by themselves.

Acknowledgments

My first and special word of gratitude goes to my wife and collaborator H. Sarbadhikari and my former teachers M. G. Nadkarni and B. V. Rao. They meticulously went through the entire script correcting numerous typos and slips as well as fine tuning the presentation. Special thanks are also due to my old friend Franco Parlamento who gave valuable suggestions, particularly on foundational issues.

I also thank Sreela Gangopadhyay, Arnab Chakraborty, Shammek Paul, and Neena Gupta for their help. I take this opportunity to thank a large number of students from all around the country including my direct students at ISI and IACS who pushed me to teach them set theory and logic year after year. This showed me that young students from all around the country are quite interested in learning foundations of mathematics for which hardly any opportunity exists in the country. This is the main motivation for me to write this book giving both the theory and the applications of set theory.

I am very grateful to Urmichhanda Bhattacharya of the Stat-Math Unit at Kolkata, ISI. I have retired from ISI more than 5 years ago. She still helped me umpteen times due to my lack of expertise in computer-related work. A major part of the book was written when both of us were confined to our homes due to COVID-19 pandemic. During this period, through internet she gave vital technical support from her home while she was busy with her household chores. I also thank my former colleague and chhatro (student) Pradipta Bandyopadhyay who helped me to solve latex-related problems.

Last but not least, I thank all my family members including my grandsons Pikku, Chikku, Totu, and Duggu for their understanding and leave me undisturbed days after days while I was writing the book.

Shashi Mohan Srivastava
Former Professor, Indian Statistical
Institute
Kolkata, India

Visiting Professor, Indian Association
for the Cultivation of Science
Kolkata, India

Ramkrishna Mission Vivekanand
Educational and Research Institute
Belur, India

Contents

1 **Introduction** 1
  Reference 7
2 **Functions and Relations** 9
  2.1 Functions 9
  2.2 Equivalence Relations 19
  2.3 Partially Ordered Sets 22
  2.4 Some More Applications of Zorn's Lemma 28
  2.5 Linearly Ordered Sets 31
  2.6 Some Historical Remarks 36
  2.7 Well-Ordered Sets 39
  2.8 Equivalence of *ZL*, *WOP*, and *AC* 45
  References 48
3 **More on Cardinals and Ordinals** 49
  3.1 Ordinal Numbers—An Informal Introduction 49
  3.2 Making Concepts Precise 50
  3.3 Cardinal Numbers 55
  3.4 An Alternative Approach to Cardinal Numbers 60
  3.5 Cardinal Arithmetic and Axiom of Choice 61
  3.6 More on Cardinals and Ordinals 62
  References 65
4 **Applications in Other Branches of Mathematics** 67
  4.1 Analysis 67
  4.2 Topology 77
  4.3 Linear Algebra 85
  4.4 Algebra 89
  4.5 Measure Theory 94
  References 109

5 Banach–Tarski Paradox 111
5.1 Hausdorff's Theorem 112
5.2 Banach–Tarski Paradox 118
References 120

6 Preliminary Concepts and Terminologies from Logic 121
References 127

7 Zermelo–Fraenkel Set Theory 129

Index 133

Chapter 1
Introduction

The purpose of this chapter is to give a flavor of axiomatic set theory and to give an idea of how to make things precise.

Like the theory of groups or rings or fields or ordered fields, etc. set theory is also a mathematical theory and an interesting one. Each conventional theory starts with some undefined concepts. For example, the group theory starts with the group operation and the group identity, and the theory of ordered fields starts with undefined addition, multiplication, additive identity, multiplicative identity, and an order relation. Similarly, set theory starts with the "belongs to" relation, denoted by $\in$.

Next, each theory declares some axioms or initial assumptions about the structural properties of the objects that it intends to study. For instance, axioms of group theory with $\cdot$ denoting the group operation and e the group identity are the following:

1. $\forall x \forall y \forall z (x \cdot (y \cdot z) = (x \cdot y) \cdot z)$.
2. $\forall x \exists y (x \cdot y = e \wedge y \cdot x = e)$.
3. $\forall x (x \cdot e = x \wedge e \cdot x = x)$.

For the sake of unambiguity and precision, it is important that each statement be "formally" expressible as above. It is very important to note that variables x, y, z, etc. that we use to formally express an axiom stand for arbitrary elements of any model of the theory (in this case any arbitrary group) and nothing else. We refer the interested reader to Chap. 6 or [1] to know what all these precisely mean.

Likewise set theory also starts with axioms. The system of axioms that we shall be working with is called *Zermelo–Fraenkel Set Theory*, in symbol ZF. All the Zermelo–Franenkel axioms are formally stated in Chap. 7. The reader may also see [1] if any confusion arises. If $x \in y$, we say that x belongs to y or x is an element of y.

The objects of the ideal model of ZF, also called *the universe*, are sets and sets alone. This means that if $x \in y$, then x is also a set. Informally, we say that elements

© The Author(s), under exclusive license to Springer Nature Singapore Pte Ltd. 2024
S. M. Srivastava, *An Introduction to Naïve Set Theory and its Applications*, University Texts in the Mathematical Sciences, https://doi.org/10.1007/978-981-97-4643-9_1

of sets are sets and nothing else. But this is a serious assumption. How do we formally express it? The following formal statement of ZF implies this:

(Extensionality Axiom.)

$$\forall x \forall y (\forall z (z \in x \leftrightarrow z \in y) \to x = y).$$

If one reads it carefully, it says that two sets x and y are equal if they contain the same <u>sets</u> z. An example here will be very enlightening. Consider the collections of all boys of a class and that of all girls in the class. Are these two collections sets? Note that these two collections say A and B, respectively, contain the same sets because none of them contain any set. But A and B are not the same collections. Hence, A and B do not satisfy the extensionality axiom and so are not sets.

What is the justification of not accepting collections like A and B, all whose elements are not sets, as sets? The main purpose of ZF is to lay the foundations of mathematics. Collections like A and B do not appear in any serious/interesting mathematics. Further, sets satisfying Zermelo–Fraenkel axioms are sufficient to define all interesting mathematical objects such as natural numbers, real numbers, complex numbers, and so on. Developing the theory of sets whose objects may not be sets also causes some logical complications that are not relevant to mathematics.

Having said this, in the Platonic world, the collections such as A and B above are regarded as sets. For example, the sample space of an experiment in statistics and many other sciences often have objects whose elements are not sets; or the domain of important random variables may contain elements which are not sets, etc. We can consider unions, intersections, functions from one such to other, etc. They can be handled following some basic principles (axioms) of ZFC. This is important and cannot be ignored. However, if we consider such sets, there is no first-order theory axiomatizing them. Since ZFC is strong enough to lay down the foundations of mathematics and stands on firm logical footing, we restrict ourselves to those sets whose elements are also sets.

It is clear that there is no difference between a set and a family (set) of sets. But we shall be better off using the familiar language of naive set theory and denote a family of sets, say as $\{A_i : i \in I\}$. In fact, from next chapter onwards, we shall only use the familiar language of naive set theory. The formal language is not that important for mathematics. However, it is crucial for axiomatic set theory and for making things precise in case of ambiguities.

Note that in conventional mathematical theories, it is easy to give models of the theory. (In the parlance of logic examples of groups, rings, fields are called models of Group Theory, Ring Theory, Field Theory respectively.) But there are serious and very deep issues involved in giving models of ZF. This is beyond the scope of this note.

1 Introduction

For this reason, an axiom of ZF asserts that *"there is a set."* Probably for the last time we show you how to express it formally.

(**Set Existence.**) $\exists x(x = x)$.

This is read as there exists a set x which is equal to itself. Since every set is equal to itself, we are only saying that there exists a set.

We stop by pointing out that *entire mathematics rests on the assumption that there is a model of ZF—in the parlance of logic this means that mathematicians assume that ZF is consistent.* As one of the finest results proved in mathematics, Gödel showed that the consistency of ZF cannot be proved in ZF, i.e., it cannot be deduced from the axioms of ZF.

Let A and B be two sets. We say that A is a *subset* of B and write $A \subset B$ or that B is a *superset* of A and write $B \supset A$, if every element of A is an element of B.

We are now in a position to give an example of a set. Take a set A (which exists by the set existence axiom) and consider

$$B = \{x \in A : x \neq x\}.$$

Since there is no $x \in A$ that is not equal to itself $B \subset A$ has no element. By extensionality axiom any two such sets are equal. We call this set the *empty set* and denote it by $\emptyset$.

Another important axiom, called *"the axiom of infinity,"* asserts that there exists an infinite set. To state it somewhat precisely, for any set x, put

$$s(x) = x \cup \{x\}.$$

There is an axiom of set theory called *the pairing axiom* from which it follows that for any set x, $\{x\}$—the collection consisting of x alone, is a set. An axiom called *the union axiom* implies that given a family of sets $\{A_i : i \in I\}$, its union consisting of those y that belongs to some $A_i, i \in I$, is a set. Now it follows that for every set x, $s(x)$ is a set.

A very interesting question arises: Can $s(x) = x$ for some x? This can be the case if and only if $x \in x$. There is an axiom of ZF, called *the foundation axiom*, which we shall discuss later, implying that for no set x, $x \in x$. Hence $s(x)$ is a proper superset of x, i.e., $s(x)$ is a superset of x but they are not equal.

We call a set A an *inductive set* if $\emptyset \in A$ and whenever $x \in A$ so does $s(x)$. It is not hard to see that if an inductive set exists it is "infinite".

(**Axiom of infinity.**) *There is an inductive set.*

Exercise 1.0.1 Express "there is an inductive set" as a formal sentence.

Take an inductive set x and consider the intersection of all inductive subsets of x. Note that this intersection is also an inductive set.

Exercise 1.0.2 Show that for all inductive sets x, the smallest inductive subset of x are all equal.

In mathematics, we denote this set by $\mathbb{N}$. Further, we set $0 = \emptyset$, $1 = s(\emptyset)$, $2 = s(1)$, and so on and note that these constitute all the element of $\mathbb{N}$. Elements of $\mathbb{N}$ are called *natural numbers*.

Remark 1.0.3 The famous mathematician Kronecker once said, "God built 1, 2, ..., the rest was built by mankind." Human beings do not know how numbers 1, 2, ... came into existence. However, starting from natural numbers, one can define integers, rational numbers, real numbers, etc. In other words, mathematics can be developed from natural numbers. Using sets we have succeeded in building natural numbers. This is one of the reasons that set theory is also called the foundation of mathematics. It would now be justified to say that "starting with sets, the rest can be built."

Cantor developed set theory non-axiomatically and somewhat naively. Some paradoxes were discovered which showed some fundamental problems in Cantor's naive set theory. The most famous such paradox, given by Russell, led to the axiomatic development of set theory. We proceed to discuss this now.

By some examples given so far, it may be clear what we mean by expressing a statement in a theory formally. The reader may see Chap. 6 or [1] for the precise definition.

Let $\varphi(x, y_0, \ldots, y_{n-1})$ be a formula of ZF. Let $b_0, \ldots, b_{n-1}$ be sets. The collection of all sets a such that $\varphi(a, b_0, \ldots, b_{n-1})$ holds, i.e., is provable using axioms of ZF, is called a *class*. In notation, we write such a class as

$$\{a : \varphi(a, b_0, \ldots, b_{n-1})\}.$$

Such a collection may or may not be a set.

Example 1.0.4 Let $\varphi(x, y)$ be the formula $x \in y$. For any set A, the collection of all a such that $a \in A$,

$$\{a \in A : \varphi(a, A)\},$$

is precisely the set A.

The above example shows that *every set is a class*. But there are classes that are not sets.

Example 1.0.5 Abbreviate the formula $\neg(x \in x)$ by $x \notin x$ and consider the following class of sets:

$$A = \{x : x \notin x\}.$$

If A were a set then it can be easily argued that

$$A \in A \Leftrightarrow A \notin A$$

which is a contradiction.

1 Introduction

Remark 1.0.6 According to Cantor's naive definition collections such as A in the last example are sets. However, assuming that A is a set, we get a contradiction. Such observations are called paradoxes. The problem lays in the non-axiomatic development of set theory by Cantor. Paradoxes disappeared with the axiomatizing of set theory and concluding that classes such as A above are not sets.

To prevent Russell's paradox, for each formula $\varphi(x, y_0, \ldots, y_{n-1})$, we give an axiom called *comprehension axiom* or sometimes *subset axiom* which we shall state informally only. (See Chap. 7.)

Subset Axiom Schema. Let $\varphi(x, y_0, \ldots, y_{n-1})$ be a formula of ZF and A a set. Let $b_0, \ldots, b_{n-1}$ be sets. Then the collection of all $a \in A$ such that $\varphi(a, b_0, \ldots, b_{n-1})$ holds is a set.

By the subset axiom schema, it follows that *the intersection of a set and a class is a set*. Note that in the proof of the existence of the empty set, the collection B considered there is a set by the subset axiom.

The class, say V, defined by the formula $x = x$ is clearly the universe of all sets.

Theorem 1.0.7 (Russell) *The class V of all sets is not a set.*

Proof If possible, suppose V is a set. Then by the subset axiom,

$$A_0 = \{x \in V : x \notin x\}$$

is a set. As seen above

$$A_0 \in A_0 \Leftrightarrow A_0 \notin A_0.$$

This contradiction proves that V is not a set. ∎

Exercise 1.0.8 Show that $\mathbb{N}$ is a set.

We give two more examples of classes that are not sets.
The ordered pair (x, y) of sets x and y is the set $\{x, \{x, y\}\}$.

Exercise 1.0.9 For sets x, y, a, b show that

$$(x, y) = (a, b) \Leftrightarrow x = a \wedge y = b.$$

Now consider the following formula $OP(x)$

$$\forall y (y \in x \leftrightarrow \exists u \exists v (y = u \vee \forall t (t \in y \leftrightarrow (t = u \vee t = v)))).$$

A careful reading of the above formula shows that for a set a, $OP(a)$ holds if and only if a is an ordered pair. We omit the proof of the fact that the class of all ordered pairs is not a set.

A set f is called a *function* if all its elements are ordered pairs and for each set a there is at most one set b such that that $(a, b) \in f$. Note that $\emptyset$ is a function called *the empty function*.

Now consider the following formula $func(x)$

$$\forall y(y \in x \to OP(y)) \land \forall u \forall v \forall w(((u, v) \in x \land (u, w) \in x) \to v = w).$$

It is clear that for a set f, $func(f)$ holds if and only if f is a function. Again we omit the proof of the fact that the class of all functions is not a set.

The most fundamental property of a class is that even if it is not a set the criterion for a set to belong to the class is captured by a formula of ZF. This allows us to handle classes rigorously. For example, the collection of all groups is not a set. But we have a definition of a group that can be expressed by a formula of ZF. It will be quite a task to write down a formula expressing that a set is a group. It is unnecessary too. However, in principle it should be possible.

We close this section by giving an axiom of set theory which is of fundamental importance to set theory. It is also called the **regularity axiom**. It is not used by mathematicians. This is the most unintuitive axiom.

Foundation. It is the following formula:

$$\forall x(\exists y(y \in x) \to \exists y(y \in x \land \neg \exists z(z \in x \land z \in y))).$$

It says that *the binary relation $\in$ is well-founded on every nonempty set*.

It has the following simple consequences whose proof we leave for the reader as as a simple exercise.

1. There is no sequence of sets $\{x_n\}$ such that for all n, $x_{n+1} \in x_n$.
2. There is no set x such that $x \in x$.

For $n \in \mathbb{N}$, we define the successor of a natural number by

$$S(n) = n + 1 = s(n).$$

Exercise 1.0.10 Show that $\mathbb{N}$ satisfies the Peano axioms.

We refer the reader to [1] for Peano axioms. $\mathbb{N}$ is also called the *standard model* of the Peano arithmetic. Since mathematics begins from the Peano arithmetic, we can see that set theory lays down the foundations of mathematics.

Unless otherwise stated, all the proofs given in this note use axioms of ZF with or without an axiom called the axiom of choice.

We have assumed rudimentary concepts in set theory like union, intersection, complement, etc. and well-known results like distributive law, De Morgan's theorem, etc. These are now part of folklore known even to senior school students in mathematics.

Reference

1. S.M. Srivastava, *A Course on Mathematical Logic*, 2nd edn. (Universitext, Springer, 2013)

Chapter 2
Functions and Relations

2.1 Functions

Let f be a function. Set
$$A = \{a : \exists b((a,b) \in f)\}.$$

We call A the *domain* of f. Since by the definition of a function, for each $a \in A$ there is a unique b such that $(a,b) \in f$, we write $f(a)$ for b. The set $\{f(a) : a \in A\}$ is called the *range* of f. Mainly using the replacement axiom of ZF it can be shown easily that the domain and the range of a function are sets. *From now on, we shall take such facts for granted without mention.* If B is any set containing the range of f, we write $f : A \to B$ and say that f is a function or a map from A to B.

Let $f : A \to B$ be a function and $C \subset A$. We define $g : C \to B$ by setting $g(a) = f(a), a \in C$. Then g is a function from C to B. We call g a *restriction* of f to C and f an *extension* of g. We denote g by $f|C$.

Let $f : A \to B$ and $g : B \to C$ be functions. For $a \in A$, we set $(g \circ f)(a) = g(f(a)) \in C$. It is entirely trivial to see that $g \circ f : A \to C$ is a function. We call $g \circ f$ the *composition* of f and g. We call a function $f : A \to B$ *one-to-one* or an *injection* if for $a_1 \neq a_2 \in A$, $f(a_1) \neq f(a_2)$. $f : A \to B$ is called *onto* B or a *surjection* if B is the range of f, i.e., for every $b \in B$ there is an $a \in A$ such that $f(a) = b$. If a map $f : A \to B$ is both one-to-one and onto B, then we call f a *bijection*. If $f : A \to B$ is a bijection, then for every $b \in B$, there is a unique $a \in A$ such that $f(a) = b$. In this case, we set $f^{-1}(b) = a$. Then $f^{-1} : B \to A$ is also a bijection and is called the *inverse* of f. For any set A, we define $id_A : A \to A$ by $id_A(a) = a, a \in A$, and call it the *identity function* on A. Note that f^{-1} is the unique function $g : B \to A$ such that $g \circ f = id_A$ and $f \circ g = id_B$. A function $f : A \to B$ such that $f(a_1) = f(a_2)$ for all $a_1, a_2 \in A$ is called a *constant* map.

The following facts are very easy to verify.

1. If $f : A \to B$ and $g : B \to C$ are injections, so is $g \circ f : A \to C$.
2. If $f : A \to B$ and $g : B \to C$ are bijections, so is $g \circ f : A \to C$.

The set of all functions $f : A \to B$ will be denoted by B^A. The product of finitely many sets $X_1, \ldots, X_n$, denoted by

$$X_1 \times \cdots \times X_n \text{ or by } \times_{i=1}^n X_i,$$

is defined to be the set

$$\{f : f : \{1, 2, \ldots, n\} \to \cup_{i=1}^n X_i \text{ such that } \forall i (f(i) \in X_i)\}.$$

The function $f : \{1, 2, \ldots, n\} \to \cup_{i=1}^n X_i$ such that $\forall i (f(i) \in X_i)$ is usually denoted by $(x_1, \ldots, x_n)$ where $x_i = f(i), i = 1, \ldots, n$, and is usually called a n-tuple.

Example 2.1.1 For any positive integer k, define $f : \mathbb{N} \to \{k, k+1, k+2, \ldots\}$ by

$$f(i) = i + k \quad i \in \mathbb{N}.$$

Then $f : \mathbb{N} \to \{k, k+1, k+2, \ldots\}$ is a bijection.

Example 2.1.2 Define $f : \mathbb{N} \to \mathbb{Z}$, the set of all integers, by

$$f(m) = \begin{cases} n & \text{if } m = 2n \\ -(n+1) & \text{if } m = 2n+1. \end{cases}$$

Then $f : \mathbb{N} \to \mathbb{Z}$ is a bijection.

Example 2.1.3 Define $f : \{0, 1\} \times \mathbb{N} \to \mathbb{N}$ by

$$f(i, n) = 2n + i, \quad i = 0, 1, n \in \mathbb{N}.$$

Then $f : \{0, 1\} \times \mathbb{N} \to \mathbb{N}$ is a bijection.

Example 2.1.4 Define $f : \mathbb{N} \times \mathbb{N} \to \mathbb{N}$ by

$$f(m, n) = 2^m(2n+1) - 1, \quad (m, n) \in \mathbb{N} \times \mathbb{N}.$$

It is easy to see that $f : \mathbb{N} \times \mathbb{N} \to \mathbb{N}$ is a bijection.

Exercise 2.1.5 Show that for every positive integer k, there is a bijection from $\mathbb{N}^{\{0,\ldots,k-1\}}$ to $\mathbb{N}$.

Exercise 2.1.6 Assume that there exist bijections between the following pair of sets:

(i) X and A, (ii) Y and B, and (iii) Z and C.

Show that there exist bijections between the following pair of sets.

1. $X \cup Y$ and $A \cup B$ provided $X \cap Y = \emptyset = A \cap B$.
2. $X \times Y$ and $A \times B$.

2.1 Functions

3. X^Y and A^B.

Exercise 2.1.7 Let X, Y and Z be sets.

1. For any $g : Z \to X^Y$ define $G : Y \times Z \to X$ by

$$G(y, z) = g(z)(y), \quad (y, z) \in Y \times Z.$$

Show that this defines a bijection from $(X^Y)^Z$ to $X^{Y \times Z}$.

2. Assume, moreover, Y and Z are disjoint sets. Show that there is a bijection between $X^{Y \cup Z}$ and $X^Y \times X^Z$.
(**Hint:** To any $g \in X^{Y \cup Z}$ assign $(g|Y, g|Z)$.)

Exercise 2.1.8 1. Let $-\infty < a < b < \infty$ and $-\infty < c < d < \infty$. Give an example of a bijection from the interval (a, b) onto the interval (c, d). If, moreover, a, b, c, d are rational numbers, give an example of a bijection $f : (a, b) \to (c, d)$ such that for any $a < x < b$, $f(x)$ is a rational number if and only if $f(x)$ is rational.

2. Show that $x \to \tan(x)$ is a bijection from $(-\frac{\pi}{2}, \frac{\pi}{2})$ onto $\mathbb{R}$.
3. If $-\infty < a < b < \infty$, show that there is a bijection between (a, b) and $\mathbb{R}$.

Exercise 2.1.9 1. Give an example of a bijection from $[0, 1)$ onto $[0, \infty)$.
2. Give an example of a bijection from $D_2 = \{z \in \mathbb{R}^2 : |z| < 1\}$ onto $\mathbb{R}^2$.
3. Give an example of a bijection from $D_n = \{z \in \mathbb{R}^n : |z| < 1\}$ onto $\mathbb{R}^n$, $n > 2$.

Exercise 2.1.10 1. Consider the map

$$(n_0, n_1, n_2, \ldots) \to \frac{1}{(n_0 + 1)+} \frac{1}{(n_1 + 1)+} \frac{1}{(n_2 + 1)+} \cdots$$

from $\mathbb{N}^\mathbb{N}$ to $(0, 1)$, where the expression on the right-hand side is a continued fraction. Show that the map is a bijection from $\mathbb{N}^\mathbb{N}$ onto the set of all irrational numbers in $(0, 1)$.

2. Show that there is a bijection from the set of all irrationals onto the set of all irrational numbers in $(0, 1)$.
(**Hint.** Take a bijection f from $\mathbb{N}$ onto $\mathbb{Z}$. Look at the sets of all irrationals in the pair of intervals $(\frac{1}{n+2}, \frac{1}{n+1})$ and $(f(n), f(n) + 1)$.)

Exercise 2.1.11 For $\alpha \in \{0, 1\}^\mathbb{N}$ and $k \in \mathbb{N}$, let $F(\alpha)(k)$ denote the number of 0s between kth. 1 and $(k + 1)$th. 1. in α. In particular, $F(\alpha)(0)$ denotes the number of 0s preceding first 1 in α. So, F is a function from $\{0, 1\}^\mathbb{N}$ to $\mathbb{N}^\mathbb{N}$. Show that $F : \{0, 1\}^\mathbb{N} \to \mathbb{N}^\mathbb{N}$ is a bijection.

Exercise 2.1.12 Let X be a set and S_X the set of all bijections from X to itself. Show that S_X is a group under the composition $\circ$ of maps.

Elements $\sigma \in S_X$ are called *permutations* of X and S_X the *permutation group* of the set X. Note that if $X = \emptyset$, then S_X consists of only one function, namely the empty function. Hence, in this case, S_X is the trivial group.

Theorem 2.1.13 (Cantor) *There is no surjection $f : \mathbb{N} \to \mathbb{R}$, where $\mathbb{R}$ is the set of all real numbers.*

Proof (Cantor) If possible, suppose f is a function from $\mathbb{N}$ onto $\mathbb{R}$. Set $a_0 = f(0)$. Let j_0 be the first natural number such that $f(j_0) > a_0$. Set $b_0 = f(j_0)$. Note that $j_0 > 0$ and for all natural number $j < j_0$, $f(j) \leq a_0$. In particular, for all such j, $f(j)$ does not belong to the open interval (a_0, b_0). Let i_0 be the first natural number such that $a_0 < f(i_0) < b_0$. Set $a_1 = f(i_0)$. Note that $i_0 > 1$ and for no $i < i_0$. $f(i) \in (a_1, b_0)$. Next let j_1 be the first natural number such that $a_1 < f(j_1) < b_0$ and put $b_1 = f(j_1)$. Then $j_1 > 2$ and for all $j < j_1$, $f(j) \notin (a_1, b_1)$.

Proceeding inductively, suppose for some natural number n,

$$a_0 < a_1 < \cdots < a_n < b_n < \cdots < b_1 < b_0$$

have been defined. Let i_{n+1} be the first natural numbers such that $a_{n+1} = f(i_{n+1}) \in (a_n, b_n)$. Next let j_{n+1} be the first natural number such that $b_{n+1} = f(j_{n+1}) \in (a_{n+1}, b_n)$. Thus, by induction, we have defined two infinite sequences of real numbers

$$a_0 < a_1 < \cdots < a_n \cdots \cdots < b_n < \cdots < b_1 < b_0$$

as above. It is not hard to show that for no natural number m, $\sup_n a_n = f(m)$. This contradiction proves our result. ∎

Remark 2.1.14 After Cantor proved these results *mankind learns for the first time that "infinite" is not one; there are infinities of different "sizes"*. Using the technique in the proof later we shall give a more general result. This is not so in case of the more popular proof using decimal expansion of real numbers.

A set Z is called *countable* if there is a surjection $f : \mathbb{N} \to Z$. Otherwise, Z is called *uncountable*. Clearly $\mathbb{N}$ is countable. Above examples show that for every positive integer k, the set $\{k, k+1, k+2, \ldots\}$, $\mathbb{Z}$, $\{0, 1\} \times \mathbb{N}$ and $\mathbb{N} \times \mathbb{N}$ are countable. By Cantor's Theorem 2.1.13, $\mathbb{R}$ is uncountable. Cantor's proof actually shows that *every non-degenerate interval of $\mathbb{R}$ is uncountable*.

Exercise 2.1.15 Show that every subset of a countable set is countable. Hence, every superset of an uncountable set is uncountable.

Exercise 2.1.16 Show that finite union of countable sets is countable.

Exercise 2.1.17 Show that the set $\mathbb{Q}$ of all rational numbers is countable.

Exercise 2.1.18 Show that the set of all irrational numbers is uncountable.

2.1 Functions

Exercise 2.1.19 Show that the set $\mathbb{N}^{<\mathbb{N}} = \cup_{k=0}^{\infty} \mathbb{N}^{\{0,1,\ldots,k-1\}}$ of all finite sequences of natural numbers is countable.

For any set X, the set of all subsets of X will be denoted by $\mathcal{P}(X)$. $\mathcal{P}(X)$ is called the *power set* of X.

Exercise 2.1.20 Show that if there is a bijection between sets X and Y, then so is between the sets $\mathcal{P}(X)$ and $\mathcal{P}(Y)$.

Theorem 2.1.21 (Cantor) *Let X be any set. Then there is no surjection $f : X \to \mathcal{P}(X)$.*

Proof If possible, suppose there is a surjection $f : X \to \mathcal{P}(X)$. Consider

$$A_0 = \{x \in X : x \notin f(x)\}.$$

Since f is a surjection, there is a $x_0 \in X$ such that $A_0 = f(x_0)$. Now we easily argue to see that

$$x_0 \in A_0 \Leftrightarrow x_0 \notin A_0.$$

This is a contradiction. ∎

Remark 2.1.22 The idea used in the above proof is extremely important. It is known as Cantor's *diagonal argument*. It may be viewed as a sort of self-reference: Think of x as a name of the subset $f(x)$ of X. Then "$x \notin f(x)$" is the statement "x does not belong to the subset whose name is x." Russell's paradox is also based on self-reference.

This technique has been adopted in a variety of situations to prove very deep results. For instance, Gödel has used this idea to prove his first incompleteness theorem; this has been used to answer the famous Halting problem in computer science and so on.

Example 2.1.23 For every $g \in \{0, 1\}^X$, put

$$F(g) = \{x \in X : g(x) = 1\}.$$

Then, $F : \{0, 1\}^X \to \mathcal{P}(X)$ is a bijection. For $A \subset X$, its *indicator* or the *characteristic* function $I_A : X \to \{0, 1\}$ is defined by

$$I_A(x) = \begin{cases} 1 \text{ if } x \in A \\ 0 \text{ if } x \in X \setminus A. \end{cases}$$

Then $A \to I_A$ from $\mathcal{P}(X)$ to $\{0, 1\}^X$ is the inverse of F.

Corollary 2.1.24 $\{0, 1\}^{\mathbb{N}}$ *is uncountable.*

The following is a very useful and interesting result.

Theorem 2.1.25 (Cantor–Dedekind Theorem) *Let X and Y be two sets such that there are injections $f : X \to Y$ and $g : Y \to X$. Then there is a bijection $h : X \to Y$.*

Proof (Dedekind) We define a map $\mathcal{H} : \mathcal{P}(X) \to \mathcal{P}(X)$ by

$$\mathcal{H}(A) = X \setminus g(Y \setminus f(A)), \quad A \subset X.$$

The following are easy to check.

(a) $A \subset B \subset X \Rightarrow \mathcal{H}(A) \subset \mathcal{H}(B)$.
(b) Since g is one-to-one, for every sequence $\{A_n\}$ of subsets of X, $\mathcal{H}(\cup_n A_n) = \cup_n \mathcal{H}(A_n)$.

Now inductively define $A_0 = \emptyset$ and $A_{n+1} = \mathcal{H}(A_n)$. Using (a) repeatedly, we see that $A_0 \subset A_1 \subset A_2 \subset \ldots$. Set $A = \cup_n A_n$. By (b), $\mathcal{H}(A) = A$. Now define $h : X \to Y$ by

$$h(x) = \begin{cases} f(x) & \text{if } x \in A \\ g^{-1}(x) & \text{if } x \in X \setminus A \end{cases}$$

Then $h : X \to Y$ is a bijection. ∎

Remark 2.1.26 The above result was conjectured and used to develop the cardinal arithmetic by Cantor. However, Cantor failed to prove this himself. He turned to Dedekind for help. Soon Dedekind sent a proof of the above result in a letter to Cantor. Dedekind's letter containing the proof got buried in papers of Cantor. It was discovered after Schröder and Bernstein independently proved the result. Later we shall see that Dedekind's proof gives a proof of a general fixed point theorem.

Exercise 2.1.27 Let $I, J \subset \mathbb{R}$ be non-degenerate intervals. Show that there is a bijection between I and J.

Let $\{X_i : i \in I\}$ be a family of sets. Their *product* is defined by

$$\times_{i \in I} X_i = \{f : f : I \to \cup_{i \in I} X_i \text{ such that } \forall i \in I (f(i) \in X_i)\}.$$

Further, assume that each $X_i \neq \emptyset$. Is it true that $\times_{i \in I} X_i \neq \emptyset$, i.e., using the axioms of ZF can we prove that the product of a family of non-empty sets is non-empty?

If I is finite, then by induction on the number of elements in the index set I, one can prove that $\times_{i \in I} X_i \neq \emptyset$. (How?) However, for I infinite, one cannot always prove it. The statement that the product of a family of non-empty sets is non-empty is known as the *axiom of choice*, *AC* in short.

Axiom of Choice. *If $\{X_i : i \in I\}$ is a family of non-empty sets, then their product $\times_{i \in I} X_i$ is non-empty.*

Apart from the fact that the axiom of choice cannot be proved from the axioms of ZF, there is also some discomfort in accepting this as an axiom for infinite I. The

2.1 Functions

statement only asserts the existence of some $f \in \times_{i \in I} X_i$ without giving any method to build one such f. A few examples will be enlightening.

Example 2.1.28 Let $f : \mathbb{R} \to \mathbb{R}$ be a function which is not continuous at 0. Then we show that there is a sequence $\{x_n\}$ of real numbers such that $x_n \to 0$ but $f(x_n) \not\to f(0)$. Usually, we prove it as follows: Since f is not continuous at 0, there is an $\epsilon > 0$ such that for every $\delta > 0$ there is a real number x with $|x| < \delta$ but $|f(x) - f(0)| \geq \epsilon$. Now for each natural number n, consider the set

$$A_n = \{x \in \mathbb{R} : |x| < \frac{1}{n+1} \wedge |f(x) - f(0)| \geq \epsilon\} \neq \emptyset.$$

By the axiom of choice, there is a sequence $\{x_n\}$ of real numbers such that for each n, $x_n \in A_n$. But we cannot precisely specify how to choose $x_n \in A_n$. So, the axiom of choice only asserts the existence of such a sequence without giving any algorithm to define it.

Example 2.1.29 Let $f : \mathbb{Q} \to \mathbb{R}$ be a function which is not continuous at 0. Then we show that there is a sequence $\{x_n\}$ of rational numbers such that $x_n \to 0$ but $f(x_n) \not\to f(0)$. Since f is not continuous at 0, there is an $\epsilon > 0$ such that for every $\delta > 0$ there is a rational number x with $|x| < \delta$ but $|f(x) - f(0)| \geq \epsilon$. Now for each natural number n, consider the set

$$A_n = \{x \in \mathbb{Q} : |x| < \frac{1}{n+1} \wedge |f(x) - f(0)| \geq \epsilon\} \neq \emptyset.$$

In one of the earlier exercises, the reader had given an example of a bijection g from $\mathbb{N}$ to $\mathbb{Q}$. For each $n \geq 0$, let i_n be the least natural number such that $g(i_n) \in A_n$ and take $x_n = g(i_n)$. Here, we have given a precise algorithm to define a sequence $\{x_n\}$ with desired properties and not used the axiom of choice.

There are some weaker forms of AC that suffice in many applications.

Countable Axiom of choice. *Let $\{A_i : i \in I\}$ be a countable family of non-empty sets. Then there is a function $f : I \to \cup_{i \in I} A_i$ such that for every $i \in I$, $f(i) \in A_i$.*

Dependent Choice, (*DC*) *Let X be a set and $A \subset X \times X$ a non-empty set such that $\pi_1(A) = X$, where π_1 is projection map to the first coordinate space. Then there exists a sequence $\{x_n\}$ in X such that for every $n \in \mathbb{N}$, $(x_n, x_{n+1}) \in A$.*

Remark 2.1.30 Since in the last two weaker versions of AC only countably many choices are made, these have wider acceptability. Most of the basic results use only countable axiom of choice or DC. For instance, DC is essentially sufficient for descriptive set theory. However, we shall not be particular and point out if in some instance we are using countable axiom of choice or DC only.

Proposition 2.1.31 *The following statements are equivalent.*

1. *AC.*
2. *Let $\{A_i : i \in I\}$ be a family of pairwise disjoint, non-empty sets. Then there is a function $f : I \to \cup_{i \in I} A_i$ such that for every $i \in I$, $f(i) \in A_i$.*
3. *Let X be a non-empty set. Then there is a function $f : \mathcal{P}(X) \setminus \{\emptyset\} \to X$ such that for every $\emptyset \neq A \subset X$, $(f(A) \in A)$.*

Proof Clearly (2) is a special case of (1). We now prove (1) assuming (2). Let $\{A_i : i \in I\}$ be a family of non-empty sets. For each $i \in I$, set $B_i = \{i\} \times A_i$. Then for $i \neq j \in I$, $B_i \cap B_j = \emptyset$. Hence, by (2), there is a function $g : I \to \cup_{i \in I} B_i$ such that for every $i \in I$, $g(i) (= (i, f(i))$, say$) \in B_i$. (2) implies (1) is clearly seen now.

(3) easily follows from (1). Assuming (3) we prove (1). Let $\{A_i : i \in I\}$ be a family of non-empty sets. Set $X = \cup_{i \in I} A_i$. By (3) there is a function $g : \mathcal{P}(X) \setminus \{\emptyset\} \to X$ such that for every $\emptyset \neq A \subset X$, $(g(A) \in A)$. Now take $f(i) = g(A_i)$, $i \in I$. ∎

Here are a few applications of AC.

A set X is called *finite* if there is a natural number n such that there is a bijection from $\{0, 1, \ldots, n - 1\}$ to X. A set X is called *infinite* if it is not finite. We first make some simple observations on finite sets.

(a) Let m, n be natural numbers and $f : \{0, 1, \ldots, m - 1\} \to \{0, 1, \ldots, n - 1\}$ an injection. Let $\{g(0), g(1), \ldots, g(m - 1)\}$ be the increasing enumeration of the range of f, i.e.,

$$g(0) = \min\{f(0), f(1), \ldots, f(m - 1)\},$$

$$g(1) = \min\{f(0), f(1), \ldots, f(m - 1)\} \setminus \{g(0)\},$$

$$g(2) = \min\{f(0), f(1), \ldots, f(m - 1)\} \setminus \{g(0), g(1)\},$$

and so on. Note that $i \leq g(i)$ for every $i < m$.

Now define h from the range of g to $\{0, 1, \ldots, n - 1\}$ by $h(g(i)) = i$, $i < m$. Thus we see that if there is a one-to-one function f from $\{0, 1, \ldots, m - 1\}$ to $\{0, 1, \ldots, n - 1\}$, then $m \leq n$. Moreover, if $m = n$, then f is onto. This proves that, "*if X is a finite set, then every injection from X to itself is a surjection. In particular, if a set Y is such that there is a one-to-one map from Y onto a proper subset of Y, then Y is infinite.*"

(b) Next assume that m, n are natural numbers and $h : \{0, 1, \ldots, n - 1\} \to \{0, 1, \ldots, m - 1\}$ is a surjection. Then $f : \{0, 1, \ldots, m - 1\} \to \{0, 1, \ldots, n - 1\}$ defined by $f(j) = \min h^{-1}(j)$, $j < m$, is an injection. Using (a) we now see, "*if X is a finite set, then every surjection from X to itself is an injection.*"

Proposition 2.1.32 *Let X be an infinite set and $x_0 \in X$. Then there is a one-to-one function from X onto $X \setminus \{x_0\}$.*

2.1 Functions

Proof By AC, there is a function $f : \mathcal{P}(X) \setminus \{\emptyset\} \to X$ such that $f(A) \in A$ for every non-empty subset A of X. By induction, we define a sequence $\{x_n\}$ in X by $x_{n+1} = f(X \setminus \{x_0, \ldots, x_n\})$. Clearly, x_n's are all distinct. Now define $h : X \to X \setminus \{x_0\}$ by

$$h(x) = \begin{cases} x_{n+1} & \text{if } x = x_n \\ x & \text{if } x \in X \setminus \{x_0, x_1, x_2, \ldots\} \end{cases}$$

Clearly $h : X \to X \setminus \{x_0\}$ is a bijection. ∎

Using induction on the number of elements in a finite set A, as a corollary we get the following result.

Proposition 2.1.33 *If X is an infinite set and $A \subset X$ a finite subset, then there is a one-to-one function from X onto $X \setminus A$.*

Summing up these observations, we see that, "*a set X is infinite if and only if there is a bijection from X onto a proper subset of itself.*"

Remark 2.1.34 Assuming AC, the above argument also proves that if X is an infinite set, then there is a one-to-one map from $\mathbb{N}$ to X.

Proposition 2.1.35 *For any two sets X and Y, the following two statements are equivalent.*

(i) There is a surjection $f : X \to Y$.
(ii) There is an injection $g : Y \to X$.

Proof Suppose there is a surjection $f : X \to Y$. Then $\{f^{-1}(y) : y \in Y\}$ is a family of non-empty sets. Hence, by AC, there is a map $g : Y \to X$ such that for every $y \in Y$, $g(y) \in f^{-1}(y)$. Since for $y_1 \neq y_2 \in Y$, $f^{-1}(y_1) \cap f^{-1}(y_2) = \emptyset$, $g(y_1) \neq g(y_2)$. Thus (i) implies (ii).

Next let $g : Y \to X$ be a one-to-one map. Take any $y_0 \in Y$. Define

$$f(x) = \begin{cases} g^{-1}(x) & \text{if } x \in g(Y) \\ y_0 & \text{if } x \notin g(Y) \end{cases}$$

Clearly $f : X \to Y$ is a surjection. ∎

Remark 2.1.36 We have used AC in proving (i) implies (ii) only. The injection $g : Y \to X$ that we obtained satisfies $f \circ g = id_Y$, the identity map on Y.

The following useful result uses countable axiom of choice only.

Proposition 2.1.37 *Let $\{A_n\}$ be a sequence of countable sets. Then $\cup_n A_n$ is countable.*

Proof For each n, fix an enumeration

$$A_n = \{a_{n0}, a_{n1}, a_{n2}, \ldots\}$$

of A_n. We have seen that there is a bijection $\alpha : \mathbb{N} \to \mathbb{N} \times \mathbb{N}$. Put $b_k = a_{\alpha(k)}$. Then $\cup_n A_n = \{b_k : k \in \mathbb{N}\}$. The result is easily seen now. ∎

Remark 2.1.38 In the last proof for each n, we have chosen an enumeration of A_n. This shows that we are using countable AC only.

Exercise 2.1.39 Show that there is a $\mathcal{A}$ of infinite subsets of $\mathbb{N}$ such that for every $A \neq B \in \mathcal{A}$, $A \cap B$ is finite and there is a bijection between $\mathcal{A}$ and $\mathbb{R}$.

(**Hint:** Fix an enumeration $\{r_n\}$ of the set $\mathbb{Q}$ of all rationals. For each irrational x, choose a sub sequence $\{r_{n_k^x}\}_{k=0}^{\infty}$ of $\{r_n\}$ such that $r_{n_k^x} \to x$ as $k \to \infty$ and consider $\{\{n_k^x\} : x \in \mathbb{R}\}$.)

Theorem 2.1.40 (D. König) *Let $\{X_i : i \in I\}$ and $\{Y_i : i \in I\}$ be two families of sets. Assume that for each $i \in I$, there is no surjection from Y_i to X_i. Then there is no surjection $f : \cup_{i \in I} Y_i \to \times_{i \in I} X_i$.*

Proof Take any map $f : \cup_{i \in I} Y_i \to \times_{i \in I} X_i$. For each $j \in I$, let $\pi_j : \times_{i \in I} X_i \to X_j$ be the projection map. By our hypothesis, for each $i \in I$, $\pi_i(f(Y_i)) \neq X_i$. Hence, $\{X_i \setminus \pi_i(f(Y_i)) : i \in I\}$ is a family of non-empty sets. Therefore, by AC, there exists a $\alpha : I \to \cup_{i \in I} X_i$ such that for each $i \in I$, $\alpha(i) \in X_i \setminus \pi_i(f(Y_i))$. Plainly $\alpha \notin f(\cup_{i \in I} Y_i)$. ∎

Exercise 2.1.41 In the last theorem, further assume that Y_i's are pairwise disjoint and for each $i \in I$, there is a one-to-one map from Y_i to X_i. Show that there is a one-to-one function from $\sum_{i \in I} Y_i$ to $\times_{i \in I} X_i$ (but not a bijection).

Let X be a non-empty set. The set of all finite sequences $(x_0, \ldots, x_{n-1})$ of elements in X including the empty sequence, which we shall denote by e, will be denoted by $X^{<\mathbb{N}}$. If $s \in X^{<\mathbb{N}}$ and $x \in X$, then $s\hat{\ }x$ will denote the *concatenation* of s and x and $|s|$ will denote the length of s. For $s, t \in X^{<\mathbb{N}}$, we write $s \prec t$ if t extends s but not equal to t.

A *tree* T on X is a non-empty set of finite sequences $(x_0, \ldots, x_{n-1})$ of elements in X such that whenever $(x_0, \ldots, x_{n-1}) \in T$, $(x_0, \ldots, x_{m-1}) \in T$ for every $m < n$. In particular, the empty sequence $e \in T$. A tree T is called *finitely splitting* if for every sequence $s \in T$, the set $\{x \in X : s\hat{\ }x \in T\}$ is finite.

Proposition 2.1.42 (J. König) *Let T be an infinite, finitely splitting tree on a set X. Then there is an infinite sequence $x_0, x_1, x_2, \ldots$ in X such that for every n, $(x_0, \ldots, x_{n-1}) \in T$.*

Proof By the hypothesis, e has infinitely many extensions in T. Suppose for some n, $x_0, \ldots, x_{n-1}$ have been defined so that $(x_0, \ldots, x_{n-1}) \in T$ and has infinitely many extensions in T. Since $\{x \in X : (x_0, \ldots, x_{n-1}, x) \in T\}$ is finite, we choose a $x_n \in X$ so that $(x_0, \ldots, x_{n-1}, x_n) \in T$ and has infinitely many extensions in T. The sequence $x_0, x_1, x_2, \ldots$ so obtained has the desired property. ∎

2.2 Equivalence Relations

Remark 2.1.43 If X were $\mathbb{N}$, we can avoid AC by choosing x_n to be the least natural with the desired property.

The following result follows from the Proposition 2.1.42 and is of use in descriptive set theory.

Proposition 2.1.44 *Let $\{A_s : s \in \mathbb{N}^{<\mathbb{N}}\}$ be a family of sets such that whenever $s \prec t$, $A_t \subset A_s$ and for every $s \in \mathbb{N}^{<\mathbb{N}}$ (including the empty sequence e),*

$$\{m \in \mathbb{N} : A_{s\frown m} \neq \emptyset\}$$

is finite. Then

$$\bigcup_{\alpha \in \mathbb{N}^{\mathbb{N}}} \bigcap_{k=0}^{\infty} A_{(\alpha(0),\ldots,\alpha(k-1))} = \bigcup_{k=0}^{\infty} \bigcup_{|s|=k} A_s.$$

Proof If an element x belongs to the set on the left-hand side, it is easy to see that it belongs to the set on the right-hand side.

Now assume that an element x belongs to the set on the right-hand side. Consider

$$T = \{s \in \mathbb{N}^{<\mathbb{N}} : x \in A_s\}.$$

By our assumptions, T is a finitely splitting tree. Since x belongs to the set on the right-hand side, T is infinite too, Hence by the Proposition 2.1.42, there is an $\alpha \in \mathbb{N}^{\mathbb{N}}$ such that for every $k \in \mathbb{N}$, $(\alpha(0),\ldots,\alpha(k-1)) \in T$. Hence, x belongs to the set on the left-hand side. ∎

Exercise 2.1.45 Let $\{A_s : s \in \mathbb{N}^{<\mathbb{N}}\}$ be a family of sets such that whenever $s \prec t$, $A_t \subset A_s$ and whenever $s \neq t \in \mathbb{N}^{<\mathbb{N}}$ and $|s| = |t|$, $A_s \cap A_t = \emptyset$. Show that

$$\bigcup_{\alpha \in \mathbb{N}^{\mathbb{N}}} \bigcap_{k=0}^{\infty} A_{(\alpha(0),\ldots,\alpha(k-1))} = \bigcup_{k=0}^{\infty} \bigcup_{|s|=k} A_s.$$

We shall be discussing the axiom of choice in greater detail later

2.2 Equivalence Relations

In this note, we shall mostly be interested in the so-called binary relations. A *binary relation* is a set all whose elements are ordered pairs. Let R be a binary relation. We define

$$\text{domain}(R) = \{a : \exists b((a,b) \in R)\},$$

$$\text{range}(R) = \{b : \exists a((a,b) \in R)\}$$

and

$$\text{field}(R) = \text{domain}(R) \cup \text{range}(R).$$

If $X \supset \text{field}(R)$, we also say that R is a binary relation on X. Thus, binary relations on X are precisely subsets of $X \times X$. It is customary to write aRb instead of $(a, b) \in R$.

Let R be a binary relation on a set X. We say that R is

1. *reflexive* if xRx for every $x \in X$;
2. *irreflexive* if $\neg(xRx)$ for every $x \in X$;
3. *symmetric* if whenever xRy holds so does yRx;
4. *asymmetric* if for every $x \neq y \in X$, $xRy \Rightarrow \neg(yRx)$;
5. *antisymmetric* if $(xRy \wedge yRx) \Rightarrow x = y$;
6. *transitive* if $(xRy \wedge yRz) \Rightarrow xRz$ and
7. *total* if for every $x, y \in X$, either xRy or yRx.

A binary relation E on X which is reflexive, symmetric, and transitive is called an *equivalence relation* on X. Let E be an equivalence relation on X. For any $x \in X$, the set

$$[x] = \{y \in X : xEy\}$$

is called an *equivalence class* or the equivalence class of x. It is easy to verify that $\{[x] : x \in X\}$ is a *partition* of X. On the other hand if Π is a partition of X, then xEy if x, y belong to the same member of the partition defines an equivalence relation on X. The set of all equivalence classes $\{[x] : x \in X\}$ is called the *quotient space* and is denoted by X/E. The map $q : X \to X/E$ defined by $q(x) = [x]$, $x \in X$, is called the *quotient map*. In fact every map on X induces an equivalence relation on X: Take any map $f : X \to Y$. Then

$$xEy \Leftrightarrow f(x) = f(y)$$

defines an equivalence relation on X.

Example 2.2.1 Let $\mathbb{N}^{\mathbb{N}}$ be the set of all functions from $\mathbb{N}$ to $\mathbb{N}$. For f and g in $\mathbb{N}^{\mathbb{N}}$, define

$$fEg \Leftrightarrow \{n \in \mathbb{N} : f(n) \neq g(n)\} \text{ is finite.}$$

It is easily checked that E is an equivalence relation on $\mathbb{N}^{\mathbb{N}}$.

For sets A and B, define

$$A \triangle B = (A \setminus B) \cup (B \setminus A) = (A \cup B) \setminus (A \cap B).$$

$A \triangle B$ is called the *symmetric difference* of A and B. We have been using parentheses to avoid confusion. For sets $A_1, A_2, \ldots, A_n$, $A_1 \triangle \cdots \triangle A_n$ is obtained by repeatedly applying $\triangle$ on $A_1, A_2, \ldots, A_n$ with the convention that parantheses are put with association to the right, For instance, $A \triangle B \triangle C = A \triangle (B \triangle C)$.

2.2 Equivalence Relations

Exercise 2.2.2 For sets A and B, show the following.

1. $A \triangle B = B \triangle A$.
2. $A \triangle \emptyset = A$.
3. $A \triangle A = \emptyset$.

Exercise 2.2.3 For sets $A_1, A_2, \ldots, A_n$, show that $A_1 \triangle A_2 \triangle \cdots \triangle A_n$ equals the set of all $x \in \cup_{i=1}^n A_i$ such that the number of i with $x \in A_i$ is odd. In particular, $\triangle$ is associative and commutative. For any set X, show that $(\mathcal{P}(X), \triangle)$ is an abelian group.

Exercise 2.2.4 For $A, B \subset X$, say AEB if $A \triangle B$ is finite. Show that E is an equivalence relation on $\mathcal{P}(X)$.

Let R be an equivalence relation on a non-empty set X. We call $A \subset X$ *invariant with respect to* R or simply *invariant* when R is understood if whenever $x \in A$ and xRy, $y \in A$. Note that A is invariant if and only if it is a union of equivalence classes. For $A \subset X$, A^* will denote the smallest invariant subset of X containing A. It is easy to check that
$$A^* = \cup\{[x] \in X/R : [x] \cap A \neq \emptyset\}.$$

A function $f : X \to Y$ is called *invariant with respect to* R or simply *invariant* when R is understood if whenever xRy, $f(x) = f(y)$.

Exercise 2.2.5 Let R be an equivalence relation on a non-empty set X, X/R the quotient space, $q : X \to X/R$ the quotient map and $f : X \to Y$ a function. Show the following.

1. $f : X \to Y$ is invariant if and only if there is a unique function $F : X/R \to Y$ such that $f = F \circ q$.
2. Let $f : X \to Y$ be invariant and F as in (1). Show that $F : X/R \to Y$ is a surjection if and only if f is so.
3. Let $f : X \to Y$ be invariant and F as in (1). Show that $F : X/R \to Y$ is an injection if and only if $xRy \Leftrightarrow f(x) = f(y)$.
4. Let $f : X \to Y$ and R be the equivalence relation on X induced by f. Assume that $F : X/R \to Y$ is the function satisfying $f = F \circ q$. Show that for $A \subset X/R$, $F(A) = f(q^{-1}(A))$ and for $B \subset Y$, $F^{-1}(B) = q(f^{-1}(B))$.

Exercise 2.2.6 Let R be an equivalence relation on a non-empty set X, X/R the quotient space, $q : X \to X/R$ the quotient map. Let $f : X \to X$ be a function. Show that there is a map $F : X/R \to X/R$ such that $F \circ q = q \circ f$ if and only if $xRy \Rightarrow f(x)Rf(y)$. Also show that such a F is a bijection if and only $xRy \Leftrightarrow f(x)Rf(y)$ and $f(X)^* = X$.

Exercise 2.2.7 Determine which of the following are equivalence relations on the set of all real numbers $\mathbb{R}$.

1. $xRy \Leftrightarrow x - y \in \mathbb{Z}$, the set of all integers.
2. $xRy \Leftrightarrow x - y \in \mathbb{Q}$, the set of all rational numbers.
3. $xRy \Leftrightarrow x - y$ is an irrational number.

Exercise 2.2.8 Let $m > 1$ be an integer. For integers x, y, define xRy if $x - y$ is divisible by m. Show that R is an equivalence relation on $\mathbb{Z}$ and $\mathbb{Z}/R$ is a finite set. Determine the number of elements in $\mathbb{Z}/R$.

Let $(G, \cdot)$ be a group and $e \in G$ the group identity. We say that G *acts* on a nonempty set X if there is a function $\cdot : G \times X \to X$, with $\cdot(g, x)$ also written as $g \cdot x$, such that for every $x \in X$ and every $g, h \in G$ the following conditions are satisfied.

1. $e \cdot x = x$.
2. $g \cdot (h \cdot x) = (g \cdot h).x$.

In this case, we also call $\cdot : G \times X \to X$ an action of G on X. Let a group G act on a set X. Note that for every $g \in G$, $x \to g \cdot x$ is a bijection with $x \to g^{-1} \cdot x$ its inverse. For $x, y \in X$, define

$$x \sim_G y \text{ if } \exists g \in G(y = g \cdot x).$$

By (1), $\sim_G$ is reflexive. Since for $x, y \in X$ and $g \in G$, $y = g \cdot x \Leftrightarrow x = g^{-1} \cdot y$, $\sim_G$ is symmetric. Now take $x, y, z \in X$ and $g, h \in G$. Let $y = g \cdot x$ and $z = h \cdot y$. Then by (2), $z = (h \cdot g) \cdot x$. This shows that $\sim_G$ is transitive. Thus, we see that $\sim_G$ is an equivalence relation on X. In this case, the equivalence class $[x]$ is called the *orbit* of x. Further, $X/\sim_G$ is usually denoted by X/G and is called the *orbit space* of the action of G on X. Group action holds a central position in mathematics. However, this is beyond the scope of this note.

2.3 Partially Ordered Sets

A binary relation on a set X which is reflexive, anti-symmetric and transitive is called a *a partial order* on X. Partial orders are generally denoted by $\leq$ with or without suffixes. Sometimes we shall also write $x \geq y$ instead of $y \leq x$. A set X equipped with a partial order $\leq$ is called a *partially ordered set* or a *poset* in short. In this case we also write $(X, \leq)$ is a poset. We shall write $x < y$ or $y > x$ if $x \neq y$ and $x \leq y$ both hold. Note that $<$ is irreflexive, asymmetric and transitive. We shall call such a relation a *strict partial order* on X.

Example 2.3.1 For every non-empty set X, $\Delta = \{(x, x) : x \in X\}$ is a partial order on X. In fact it is the smallest partial order on X in the sense that every partial order on X contains Δ. Sometime Δ is called the *trivial partial order* on X.

2.3 Partially Ordered Sets

Example 2.3.2 Let X be any set. Then $(\mathcal{P}(X), \subset)$ and $(\mathcal{P}(X), \supset)$ are posets.

Let $(X, \leq)$ be a poset, $A \subset X$ and $x \in X$. We say that x is an *upper bound* (respectively a *lower bound*) of A if for every $y \in A$, $y \leq x$ (respectively $x \leq y$). If A has an upper bound (respectively a lower bound) in X, then we call A *bounded above* (respectively *bounded below*). An element x of X is called the *least upper bound/lub* (respectively *greatest lower bound/glb*) of A if x is an upper bound (respectively a lower bound) of A and for every upper bound (respectively lower bound) z of A, $x \leq z$ (respectively $z \leq x$). Note that lub/glb of A, if it exists, is unique. An element $x \in X$ is called the *maximum* (respectively *minimum*) element of X if for every $y \in X$, $y \leq x$ (respectively $x \leq y$). Note that X can have at most one maximum element and at most one minimum element. An element $x \in X$ is called a *maximal* (respectively *minimal*) element of X if there is no $y \in X$ such that $y > x$ (respectively $x > y$). A subset C of X is called a *chain* in X if for every $x, y \in C$ either $x \leq y$ or $y \leq x$, i.e., $\leq$ restricted to C is total.

Example 2.3.3 Let $X = (-1, 1] \times (-1, 1]$. For $(a, b), (c, d) \in X$, set

$$(a, b) \leq (c, d) \Leftrightarrow (a \leq c \wedge b \leq d).$$

Then $(1, 1)$ is the maximum element of X while X has no minimum element.

Example 2.3.4 Let $X = (\{0\} \times ((-1, -1/2) \cup (0, 1])) \cup ([-1, 1] \times \{0\})$. For $(a, b), (c, d) \in X$ define

$$(a, b) \leq (c, d) \Leftrightarrow (a \leq c \wedge b \leq d).$$

Take $A = \{0\} \times (0, 1/2)$. Then A is both bounded above and bounded below. However, it has a lub $(0, 1/2)$ but no glb. $(0, 1)$ and $(1, 0)$ are the two maximal elements of X. $(-1, 0)$ is the only minimal element of X but it is not the minimum element of X.

The most important statement concerning posets is known as *Zorn's Lemma* (ZL in short).

Zorn's Lemma. *Let $(\mathbb{P}, \leq)$ be a non-empty partially ordered set such that every chain in $\mathbb{P}$ has an upper bound. Then $\mathbb{P}$ contains a maximal element.*

Zorn's lemma is very widely used in almost all branches of mathematics. However, in this chapter, we shall give its applications in set theory only. We shall give many applications of ZL in mathematics later. Its status in set theory will be discussed in detail later.

Proposition 2.3.5 *The following statements are equivalent.*

1. *ZL.*
2. *Every non-empty poset contains a maximal chain (i.e., a chain that is not properly contained in a chain).*

Proof Assume ZL and let $(\mathbb{P}, \leq)$ be a non-empty poset. Since for any $x \in \mathbb{P}$, $\{x\}$ is a chain in $\mathbb{P}$, the set $\mathcal{C}$ of all chains in $\mathbb{P}$ is a non-empty set. Now consider the non-empty poset $(\mathcal{C}, \subset)$. If $\{C_i : i \in I\}$ is a chain in $\mathcal{C}$, then $\cup_{i \in I} C_i$ is a chain which is an upper bound of $\{C_i : i \in I\}$. Hence, $\mathbb{P}$ has a maximal chain by ZL.

Conversely, let $(\mathbb{P}, \leq)$ be a non-empty poset such that every chain in $\mathbb{P}$ has an upper bound. By (2), $\mathbb{P}$ has a maximal chain, say C. By the hypothesis, C has an upper bound which is a maximal element of $\mathbb{P}$ because C is a maximal chain. ∎

Proposition 2.3.6 *ZL implies AC.*

Proof Let $\{A_i : i \in I\}$ be a family of non-empty sets. Define

$$\mathbb{P} = \{(J, f) : J \subset I \ \& \ f : J \to \cup_{j \in J} A_j \text{ such that } \forall j \in J (f(j) \in A_j)\}.$$

For $(J_1, f_1), (J_2, f_2) \in \mathbb{P}$, set

$$(J_1, f_1) \leq (J_2, f_2) \Leftrightarrow J_1 \subset J_2 \ \& \ f_2|J_1 = f_1,$$

where $f_2|J_1$ denotes the restriction of f_2 to J_1.

The empty function belongs to $\mathbb{P}$ implying that $\mathbb{P}$ is non-empty. It is fairly routine to check that $\leq$ is a partial order on $\mathbb{P}$.

Let $C = \{(J_a, f_a) \in \mathbb{P} : a \in A\}$ be a chain in $\mathbb{P}$. Then this is a consistent set of functions. Take $J = \cup_{a \in A} J_a$ and $f = \cup_{a \in A} f_a$, the smallest common extension of each of $f_a, a \in A$. Clearly $(J, f) \in \mathbb{P}$ is an upper bound of C. Hence, by ZL, $\mathbb{P}$ has a maximal element, say (J_0, f_0). Our proof will be complete, if we show that $J_0 = I$.

Suppose this is not the case. Take any $i \in I \setminus J_0$. Since $A_i \neq \emptyset$, there is a $a_i \in A_i$. Set $g = f_0 \cup \{(i, a_i)\}$. Then $(J_0 \cup \{i\}, g) > (J_0, f_0)$, contradicting the maximality of (J_0, f_0). ∎

Proposition 2.3.7 (Teichmüller–Tukey Lemma) *Let $\mathcal{A}$ be a non-empty set such that for every finite set B,*

$$B \in \mathcal{A} \Leftrightarrow \text{ every finite subset of } B \text{ belongs to } \mathcal{A}.$$

Show that $\mathcal{A}$ contains a maximal set.

Proof Consider the non-empty poset $\mathbb{P} = (\mathcal{A}, \subset)$. Let $\mathcal{C} = \{A_i : i \in I\}$ be a chain in $\mathbb{P}$. This means that for all $i, j \in I$, either $A_i \subset A_j$ or $A_j \subset A_i$. Set $A = \cup \mathcal{C}$. Let $B \subset A$ be finite. Since $\mathcal{C}$ is a chain, there is a $i \in I$ such that $B \subset A_i$. Since $A_i \in \mathcal{A}$, $B \in \mathcal{A}$. Hence, $A \in \mathcal{A}$ by our hypothesis. Thus, A is an upper bound of $\mathcal{C}$. The result follows from Zorn's lemma. ∎

Remark 2.3.8 Like AC, ZL merely asserts the existence of a maximal element without giving any example of such an element or without giving a method of constructing such a set.

2.3 Partially Ordered Sets

Let X be a non-empty set and $\mathcal{F}$ a family of subsets of X. We say that $\mathcal{F}$ has *finite intersection property* (*f.i.p.* in short) if for every finite $F_1, \ldots, F_n \in \mathcal{F}$, $\cap_{i=1}^{n} F_i \neq \emptyset$. For example, let $X = \mathbb{N}$ and $\mathcal{F} = \{\{n, n+1, n+2, \ldots\} : n \in \mathbb{N}\}$. Then $\mathcal{F}$ has f.i.p. We say that a non-empty $\mathcal{F} \subset \mathcal{P}(X)$ is a *filter* on X if it satisfies the following conditions:

1. $\emptyset \notin \mathcal{F}$.
2. $A, B \in \mathcal{F} \Rightarrow A \cap B \in \mathcal{F}$.
3. If $A \in \mathcal{F}$ and $A \subset B \subset X$, then $B \in \mathcal{F}$.

Note that if $\mathcal{F}$ is a filter on X, then $X \in \mathcal{F}$, $\mathcal{F}$ has f.i.p. and is closed under finite intersections.

If $\emptyset \neq A \subset X$, then $\mathcal{F} = \{B \subset X : A \subset B\}$ is a filter on X. A subset A of X is called *co-finite* if its complement $X \setminus A$ is finite. The family $\mathcal{F}$ of all co-finite subsets of $\mathbb{N}$ is a filter on $\mathbb{N}$.

Lemma 2.3.9 *Every non-empty family of subsets of a set X with f.i.p. is contained in a filter on X.*

Proof Let $\mathcal{B}$ be a non-empty family of subsets of X with f.i.p. Define

$$\mathcal{F} = \{A \subset X : \exists n \geq 1 \exists B_1, \ldots, B_n \in \mathcal{B} (A \supset \cap_{i=1}^{n} B_i)\}.$$

It is quite routine to check that $\mathcal{F}$ is a filter on X containing $\mathcal{B}$. ∎

Remark 2.3.10 $\mathcal{F}$ is the smallest filter on X containing $\mathcal{B}$. It is also called the filter generated by $\mathcal{B}$.

A family of subsets of a set with f.i.p. is also called a *filter base*.

Let $\mathbb{P}$ denote the set of all filters on a set X. This is clearly a non-empty set and $(\mathbb{P}, \subset)$ is a poset. A maximal element of this poset is called a *maximal filter* or an *ultrafilter* on X. So an ultrafilter $\mathcal{U}$ on a set X is a filter on X such that there is no filter on X containing $\mathcal{U}$ properly.

Example 2.3.11 Let X be a non-empty set. Then for each $x \in X$,

$$\mathcal{U}_x = \{A \subset X : x \in A\}$$

is an ultrafilter on X. That $\mathcal{U}_x$ is a filter is clear. To see that it is a maximal filter, suppose to the contrary. Get a filter $\mathcal{F}$ on X containing $\mathcal{U}_x$ properly. Let $A \in \mathcal{F} \setminus \mathcal{U}_x$. Then $x \notin A$ and both A and $\{x\}$ belong to $\mathcal{F}$. Hence, $\emptyset = A \cap \{x\} \in \mathcal{F}$ which is a contradiction.

Exercise 2.3.12 Let $\mathcal{U}$ be an ultrafilter on a non-empty set X. Assume that $A = \cap \mathcal{U}$ is non-empty. Show that A is a singleton.

Proposition 2.3.13 *Every filter on a non-empty set is contained in a maximal filter on X.*

Proof Let $\mathcal{F}$ be a filter on a non-empty set X. Consider

$$\mathbb{P} = \{\mathcal{F}' : \mathcal{F}' \supset \mathcal{F} \text{ is a filter on } X\}.$$

Since $\mathcal{F} \in \mathbb{P}, \mathbb{P} \neq \emptyset$. Clearly, $(\mathbb{P}, \subset)$ is a poset. Let $\mathcal{C}$ be a chain in $\mathbb{P}$. Then, for every $\mathcal{F}', \mathcal{F}'' \in \mathcal{C}$, either $\mathcal{F}' \subset \mathcal{F}''$ or $\mathcal{F}'' \subset \mathcal{F}'$. Consider

$$\mathcal{G} = \{A \subset X : \exists \mathcal{F}' \in \mathcal{C}(A \in \mathcal{F}')\}.$$

It is not hard to check that $\mathcal{G}$ is a filter on X that contains every filter in $\mathcal{C}$. Thus we see that every chain in $\mathbb{P}$ has an upper bound. Hence, by ZL, there is a maximal filter on X containing $\mathcal{F}$. ∎

Remark 2.3.14 Earlier we have seen that the set of all co-finite subsets of $\mathbb{N}$ is a filter on $\mathbb{N}$. Indeed, let X be an infinite set. Then the set of all co-finite subsets of X is a filter on X. Hence, by the last result, there is an ultrafilter $\mathcal{U}$ on X that contains all cofinite sets. For any such ultrafilter $\cap \mathcal{U} = \emptyset$. On the other hand, assume that $\mathcal{U}$ is an ultrafilter on X such that $\cap \mathcal{U} \neq \emptyset$. Then we have seen that $\mathcal{U} = \mathcal{U}_x$ for some $x \in X$. An ultrafilter of the form $\mathcal{U}_x$ is called a *principal ultrafilter* at x. Other ultrafilters are called *free ultrafilters*.

Exercise 2.3.15 Let $\mathcal{U}$ be an ultrafilter on a non-empty set that contains a finite set. Show that $\mathcal{U}$ is a principal ultrafilter.

Theorem 2.3.16 *Let $\mathcal{F}$ be a filter on X. The following statements are equivalent.*

(i) $\mathcal{F}$ is a maximal filter.
(ii) For every $A \subset X, \forall B \in \mathcal{F}(B \cap A \neq \emptyset) \Rightarrow A \in \mathcal{F}$.
(iii) Whenever $A_1 \cup \cdots \cup A_n \in \mathcal{F}$, $A_i \in \mathcal{F}$ for some $1 \leq i \leq n$.
(iv) For every $A \subset X$, exactly one of A and $X \setminus A$ is in $\mathcal{F}$.

Proof Assume that $\mathcal{F}$ is a maximal filter and $A \subset X$ is such that $A \cap B \neq \emptyset$ for every $B \in \mathcal{F}$. Then $\mathcal{F} \cup \{A\}$ has f.i.p. Therefore, by the lemma, it is contained in a filter, say $\mathcal{F}'$. Since $\mathcal{F}$ is a maximal filter, we have $\mathcal{F} = \mathcal{F}'$. This implies that $A \in \mathcal{F}$. Thus, (i) implies (ii) follows.

Next assume (ii). If possible, suppose there exists $\cup_{i=1}^{n} A_i \in \mathcal{F}$ such that none of $A_1, \ldots, A_n$ belong to $\mathcal{F}$. By (ii), for each $1 \leq i \leq n$, there exists a $B_i \in \mathcal{F}$ such that $A_i \cap B_i = \emptyset$. Since $\mathcal{F}$ is a filter, $\cap_{i=1}^{n} B_i \in \mathcal{F}$. This implies that $\emptyset = (\cup_{i=1}^{n} A_i) \cap (\cap_{i=1}^{n} B_i) \in \mathcal{F}$. This contradiction proves that (ii) implies (iii).

Since $X = A \cup (X \setminus A) \in \mathcal{F}$ and $\mathcal{F}$ is a filter, clearly, (iii) implies (iv).

Finally, assume that $\mathcal{F}$ is a filter but not a maximal one. Take a filter $\mathcal{F}'$ that contains $\mathcal{F}$ properly. Get $A \in \mathcal{F}' \setminus \mathcal{F}$. If possible, suppose $X \setminus A \in \mathcal{F} \subset \mathcal{F}'$. But then both A and $X \setminus A$ belong to $\mathcal{F}'$ implying that $\emptyset = A \cap (X \setminus A) \in \mathcal{F}'$ which is a contradiction. We have now proved that (iv) implies (i) and the proof of our theorem is complete. ∎

2.3 Partially Ordered Sets

Sets in a filter are regarded as large in some sense. A few examples given below should make it clear. The readers are required to have some knowledge of topology and measure theory to understand the examples.

Example 2.3.17 Let $\mathcal{F} = \{A \subset \mathbb{R} : \lambda(\mathbb{R} \setminus A) = 0\}$, where λ denotes the Lebesgue measure on $\mathbb{R}$. Sets in $\mathbb{R}$ are also called *co-null sets*. They are "large" sets with respect to the Lebesgue measure. It is quite easy to check that $\mathcal{F}$ is a filter on $\mathbb{R}$. It is not an ultrafilter because every Lebesgue measurable subset of $\mathbb{R}$ of positive measure intersects every co-null sets.

Example 2.3.18 A subset $A \subset \mathbb{R}$ is called *nowhere dense* if its closure $\overline{A}$ does not contain any non-empty open set. A subset B of $\mathbb{R}$ is called *meager* if it is a countable union of nowhere dense sets. The complement in $\mathbb{R}$ of a meager set is called a *co-meager set*. It is easy to see that the set $\mathcal{F}$ of all co-meager sets is a filter but not an ultrafilter. Sets in $\mathcal{F}$ are considered topologically large sets.

Let $\{X_a : a \in A\}$ be a non-empty family of non-empty sets and $\mathcal{F}$ a filter on A. Set $X = \times_{a \in A} X_a$. For $\alpha, \beta \in X$, define

$$\alpha \sim \beta \Leftrightarrow \{a \in A : \alpha(a) = \beta(a)\} \in \mathcal{F}.$$

It is easy to see that $\sim$ is an equivalence relation on X. The set of all $\sim$-equivalence classes is denoted by $(\times_{a \in A} X_a)/\mathcal{F}$. In case $\mathcal{U}$ is an ultrafilter on A, the set $(\times_{a \in A} X_a)/\mathcal{U}$ is called an *ultraproduct* of $\{X_a : a \in A\}$. The ultraproduct of groups, fields, models of theories etc. are very useful. We shall use this notion in the section on algebra in Chap. 4.

Exercise 2.3.19 Let $\{X_a : a \in A\}$ be a family of non-empty sets, $a_0 \in A$ and $\mathcal{U} = \{I \subset A : a_0 \in I\}$, the principal ultrafilter at a_0. Show that there is a bijection between X_{a_0} and $(\times_{a \in A} X_a)/\mathcal{U}$.

Just as sets in a filter are considered large sets in some sense, there is a dual notion of a filter, an ideal, whose members are considered small sets in some sense.

Let X be a non-empty set. An *ideal* on X is a non-empty family $\mathcal{I}$ of subsets of X satisfying the following two conditions.

1. $A, B \in \mathcal{I}$ implies $A \cup B \in \mathcal{I}$.
2. Whenever $A \in \mathcal{I}$ and $B \subset A$, $B \in \mathcal{I}$.

Note that the empty set belongs to every ideal. An ideal $\mathcal{I}$ on X is called *proper* if $\mathcal{I} \neq \mathcal{P}(X)$. Clearly $\mathcal{I}$ is a proper ideal on X if and only if $X \notin \mathcal{I}$.

Using Zorn's lemma, we can easily prove the following.

Proposition 2.3.20 *Every proper ideal on a non-empty set is contained in a maximal proper ideal.*

Its fairly simple proof is left to the reader as an exercise. An ideal $\mathcal{I}$ on a set X is called a *σ-ideal* if it is also closed under countable unions.

Example 2.3.21 Let X be an uncountable set. Then $\mathcal{I} = \{A \subset X : A \text{ countable}\}$ is a proper σ-ideal.

Example 2.3.22 $\mathcal{I} = \{A \subset \mathbb{R} : \lambda(A) = 0\}$, λ the Lebesgue measure, is a proper σ-ideal on $\mathbb{R}$.

Example 2.3.23 $\mathcal{I} = \{A \subset \mathbb{R} : A \text{ is meager}\}$ is a proper σ-ideal on $\mathbb{R}$. That $\mathbb{R} \notin \mathcal{I}$ follows from the Baire category theorem. (See [1], Theorem 2.5.5.)

2.4 Some More Applications of Zorn's Lemma

In this section, we give some applications of Zorn's lemma which are particularly useful for cardinal arithmetic.

Theorem 2.4.1 *For any two sets X and Y either there is an injection from X to Y or an injection from Y to X.*

Proof Consider

$$\mathbb{P} = \{(A, f) : A \subset X \ \& \ f : A \to Y \text{ an injection}\}.$$

Since the empty function belongs to $\mathbb{P}$, $\mathbb{P} \neq \emptyset$. For $(A, f), (B, g) \in \mathbb{P}$, define

$$(A, f) \leq (B, g) \Leftrightarrow A \subset B \ \& \ g|A = f.$$

It is easy to see that $\leq$ is a partial order on $\mathbb{P}$.

Let $\mathcal{C} = \{(A_i, f_i) : i \in I\}$ be a chain in $\mathbb{P}$. Put $A = \cup_{i \in I} A_i$ and $f : A \to Y$ be the unique map satisfying $f|A_i = f_i$ for each $i \in I$. Since f_i's are injections, $f : A \to Y$ is an injection. Clearly, (A, f) is an upper bound of $\mathcal{C}$. Hence, by ZL, $\mathbb{P}$ has a maximal element, say (A_0, f_0).

If $A_0 = X$, we see that there is an injection $f_0 : X \to Y$. If $f_0(A_0) = Y$, then $f_0^{-1} : Y \to X$ is an injection. Suppose neither is the case. Then there exist $x_0 \in X \setminus A_0$ and $y_0 \in Y \setminus f_0(A_0)$. Then $(A_0, f_0) < (A_0 \cup \{x_0\}, f_0 \cup \{(x_0, y_0)\})$ contradicting the maximality of (A_0, f_0). ∎

Theorem 2.4.2 *Let X be an infinite set. Then there is a bijection $f : X \times \{0, 1\} \to X$.*

Proof Let

$$\mathbb{P} = \{(A, f) : A \subset X \ \& \ f : A \times \{0, 1\} \to A \text{ is a bijection}\}.$$

To show that $\mathbb{P} \neq \emptyset$, either one sees that the empty function belongs to $\mathbb{P}$ or since X is infinite, it contains a countably infinite subset, say A. Since there is a bijection from $\mathbb{N} \times \{0, 1\} \to \mathbb{N}$, there is a bijection $f : A \times \{0, 1\} \to A$. Thus, $\mathbb{P} \neq \emptyset$.

2.4 Some More Applications of Zorn's Lemma

For $(A, f), (B, g) \in \mathbb{P}$, define

$$(A, f) \leq (B, g) \Leftrightarrow A \subset B \ \& \ g|A \times \{0, 1\} = f.$$

It is easily seen that $\leq$ is a partial order on $\mathbb{P}$.

Let $\mathcal{C} = \{(A_i, f_i) : i \in I\}$ be a chain in $\mathbb{P}$. Put $A = \cup_{i \in I} A_i$ and $f : A \times \{0, 1\} \to A$ be the unique map satisfying $f|A_i \times \{0, 1\} = f_i$ for each $i \in I$. It is easily argued that $(A, f) \in \mathbb{P}$ and is an upper bound of $\mathcal{C}$. Hence, by ZL, $\mathbb{P}$ has a maximal element, say (A_0, f_0).

If possible, suppose $X \setminus A_0$ is infinite. Then there is a countably infinite subset A of $X \setminus A_0$. Take a bijection $g : A \times \{0, 1\} \to A$. Then $(A_0, f_0) < (A_0 \cup A, f_0 \cup g)$, contradicting the maximality of (A_0, f_0).

So, $X \setminus A_0$ is finite. Since X is infinite, there is a bijection $h : X \to A_0$. Using this, we see that there is a bijection from $X \times \{0, 1\}$ to X. ∎

Corollary 2.4.3 *If X is an infinite set, then for every positive integer $k \geq 1$, there is a bijection from $X \times \{0, \ldots, k-1\} \to X$.*

This can now be easily proved by induction on k. These results amount to saying that

Remark 2.4.4 For every positive integer $k > 1$, every infinite set X can be partitioned into k-many subsets $X_1, \ldots, X_k$ such that there is a bijection $f_i : X_i \to X$ for each $1 \leq i \leq k$.

Corollary 2.4.5 *Let X and Y be disjoint sets with X infinite. Assume that there is an injection $f : Y \to X$. Then there is a bijection from $X \cup Y$ to X.*

Proof Clearly there is an injection from X to $X \cup Y$, namely the inclusion map $i : X \hookrightarrow X \cup Y$. Let $h : X \cup Y \to (X \times \{0\}) \cup (Y \times \{1\})$ be defined by

$$h(a) = \begin{cases} (a, 0) \text{ if } a \in X \\ (a, 1) \text{ if } a \in Y \end{cases}$$

Then h is an injection. Next define $g : (X \times \{0\}) \cup (Y \times \{1\}) \to X \times \{0, 1\}$ by

$$g(a, i) = \begin{cases} (a, 0) & \text{if } i = 0 \\ (f(a), 1) & \text{if } i = 1 \end{cases}$$

We note again that g is a injection. In Theorem 2.4.2 that there is a bijection $u : X \times \{0, 1\} \to X$. Then $u \circ g \circ h : X \cup Y \to X$ is an injection.

Our result now follows from Theorem 2.1.25. ∎

Theorem 2.4.6 *For every infinite set X, there is a bijection from $X \times X$ to X.*

Proof Let

$$\mathbb{P} = \{(A, f) : A \subset X \ \& \ f : A \times A \to A \text{ is a bijection}\}.$$

To show that $\mathbb{P} \neq \emptyset$, either one sees that the empty function belongs to $\mathbb{P}$ or since X is infinite, it contains a countably infinite subset, say A. Since there is a bijection from $\mathbb{N} \times \mathbb{N} \to \mathbb{N}$, there is a bijection $f : A \times A \to A$. Thus, $\mathbb{P} \neq \emptyset$.

For $(A, f), (B, g) \in \mathbb{P}$, define

$$(A, f) \leq (B, g) \Leftrightarrow A \subset B \ \& \ g|A \times A = f.$$

It is easily seen that $\leq$ is a partial order on $\mathbb{P}$.

Let $\mathcal{C} = \{(A_i, f_i) : i \in I\}$ be a chain in $\mathbb{P}$. Put $A = \cup_{i \in I} A_i$ and $f : A \times A \to A$ be the unique map satisfying $f|A_i \times A_i = f_i$ for each $i \in I$. It is easily argued that $(A, f) \in \mathbb{P}$ and is an upper bound of $\mathcal{C}$. Hence, by ZL, $\mathbb{P}$ has a maximal element, say (A_0, f_0).

If possible, suppose there is an injection $h : A_0 \to X \setminus A_0$. Set $B = h(A_0)$. By Remark 2.4.4, we can partition B into three disjoint sets B_1, B_2 and B_3 such that there is a bijection $g_i : B_i \to A_0$ for $i = 1, 2, 3$. Using h, g_1, g_2, g_3 and f_0, it is not hard to define bijections $h_1 : B \times A_0 \to B_1, h_2 : B \times B \to B_2$ and $h_3 : A_0 \times B \to B_3$. Let $u : (A_0 \cup B) \times (A_0 \cup B) \to A_0 \cup B$ be the map satisfying $u|A_0 \times A_0 = f_0, u|B \times A_0 = h_1, u|B \times B = h_2$ and $u|A_0 \times B = h_3$. Then $(A_0 \cup B, u) \in \mathbb{P}$ and $(A_0, f_0) < (A_0 \cup B, u)$. This contradicts the maximality of (A_0, f_0).

It follows that there is no injection from A_0 to $X \setminus A_0$. Hence, by the first result of this section, there is an injection from $X \setminus A_0$ to A_0. By the last corollary, it follows that there is a bijection between X and A_0. Since there is a bijection $f_0 : A_0 \times A_0 \to A_0$, it is now easy to conclude that there is a bijection from $X \times X$ to X. ∎

Exercise 2.4.7 Assume that X is an infinite set, Y a non-empty set and there is an injection from Y to X. Show that there is a bijection between $X \times Y$ and X.

Let X be a non-empty set. A finite sequence of elements of X is called a *word* on alphabet X. The set of all words on alphabet X will be denoted by $X^{<\mathbb{N}}$. For $s \in X^{<\mathbb{N}}$, $|s|$ will denote the length of the finite sequence s, i.e., if $s = (x_0, \ldots, x_{k-1})$ (s is the empty word, denoted by e, if $k = 0$), then $|s| = k$.

Exercise 2.4.8 Show the following.

1. If X is finite, then $X^{<\mathbb{N}}$ is countable.
2. If X is infinite, then there is a bijection from $X^{<\mathbb{N}}$ to X.

2.5 Linearly Ordered Sets

Let X be a non-empty set. A partial order on X which is total is called a *linear order* on X. In general, we shall denote a linear order by $\leq$ with or without suffix. If $\leq$ is a linear order on X, we say that $(X, \leq)$ or simply X, when there is no scope for confusion, is a *linearly ordered set* or a *loset*. Note that if a loset X has a maximal (minimal) element, it is unique and is the maximum (respectively minimum) element of X. The maximum (minimum) element of a loset X is also called the last (respectively the first) element of X. A subset A of X is called *bounded* if it is both bounded above and bounded below. For $x \neq y \in X$, we shall write $x < y$ if $x \leq y$ holds. Sometimes we call $<$ a *strict linear order* on X. We can easily check that a binary relation $<$ is a strict linear order if and only if it is irreflexive, asymmetric, transitive and the following law, called the *trichotomy law* holds. For every $x, y \in X$ exactly one of the following three relations

$$x < y \text{ or } x = y \text{ or } y < x$$

holds.

Example 2.5.1 The usual order on $\mathbb{R}$ is a linear order.

Example 2.5.2 Let $(X_1, \leq_1)$ and $(X_2, \leq_2)$ be linearly ordered sets and $X = X_1 \times X_2$. For $(x_1, y_1) \neq (x_2, y_2)$ belonging to X, define

$$(x_1, y_1) < (x_2, y_2) \Leftrightarrow \text{ either } x_1 < x_2 \text{ or } (x_1 = x_2 \ \& \ y_1 < y_2).$$

It is easy to check that $<$ is a strict linear order on X. It is often referred to as the *lexicographic order* and is denoted by $<_{\text{lex}}$. Similarly if $(X_1, \leq_1), \ldots, (X_n, \leq_n)$ are losets and $X = X_1 \times \cdots \times X_n$, we define the lexicographic order $<_{\text{lex}}$ on X as follows: for $(x_1, \ldots, x_n) \neq (y_1, \ldots, y_n) \in X$

$$(x_1, \ldots, x_n) <_{\text{lex}} (y_1, \ldots, y_n) \Leftrightarrow \exists 1 \leq i \leq n (\forall j < i (x_j = y_j) \wedge x_i <_i y_i).$$

Example 2.5.3 Let $(Y, \leq)$ be a loset and $X = Y^{\mathbb{N}}$. For $\alpha \neq \beta \in X$, define

$$\alpha <_{\text{lex}} \beta \Leftrightarrow \exists i \in \mathbb{N} (\forall j < i)(\alpha(j) = \beta(j)) \wedge \alpha(i) < \beta(i)).$$

Then $<_{\text{lex}}$ is a strict linear order on X.

Example 2.5.4 Let $(X, \leq)$ be a loset and $W = X^{<\mathbb{N}}$, the set of all words on alphabet X. For $s \neq t \in W$, define

$$s <' t \Leftrightarrow \text{ either } |s| < |t| \text{ or } (|s| = |t| \wedge s <_{\text{lex}} t).$$

Then $<'$ is a strict linear order on W.

Let $(X_1, \leq_1)$ and $(X_2, \leq_2)$ be losets. A function $f : X_1 \to X_2$ is called *order preserving* if for all $x, y \in X_1$, $x \leq_1 y \Rightarrow f(x) \leq_2 f(y)$. A one-to-one, order preserving map $f : X_1 \to X_2$ is called an *embedding* of X_1 into X_2. Note that a function $f : X_1 \to X_2$ is an embedding if and only if for every $x, y \in X_1, x <_1 y \Leftrightarrow f(x) <_2 f(y)$. An order preserving bijection $f : X_1 \to X_2$ is called an *order isomorphism* or simply an *isomorphism*. Two losets X_1 and X_2 are called *order isomorphic* or simply *isomorphic* if there is an isomorphism from one to the other. A loset $(X, \leq)$ is called *order dense* if for all $x < y$ in X there is a $z \in X$ such that $x < z < y$. Note that every order dense linearly ordered dense set with more than one point is infinite. A loset X is said to be *complete* or is said to satisfy *least upper bound axiom* (*lub axiom* in short) if every bounded above subset A of X has a least upper bound in X.

Let X be a loset and $a, b \in X$. We define

$$(-\infty, a) = \{x \in X : x < a\},$$

$$(b, \infty) = \{x \in X : b < x\}$$

and if $a < b$, then

$$(a, b) = \{x \in X : a < x < b\}.$$

Such sets are called *open intervals*. The loset $(X, \leq)$ is called *separable* if there is a countable $D \subset X$ such that every non-empty open interval of X contains a point of D. Such a set D is called *dense* in X. The loset X is said to satisfy *countable chain condition* or *to have c.c.c.* if every family of pairwise disjoint non-empty open intervals in X is countable.

Lemma 2.5.5 *Every separable linearly ordered set satisfies countable chain condition.*

Proof Let X be a separable loset and $D \subset X$ a countable dense set. Let $\mathcal{I}$ be a family of pairwise disjoint, non-empty open intervals in X. By AC, for every $I \in \mathcal{I}$, there is a point $f(I) \in I \cap D$. Since intervals in $\mathcal{I}$ are pairwise disjoint, $f : \mathcal{I} \to D$ is one-to-one. Since D is countable, it follows that $\mathcal{I}$ is countable. ∎

A subset U of X is called *open* if it is a union of open intervals. The reader familiar with point set topology would notice that if $X = \mathbb{R}$ this is precisely the usual topology of $\mathbb{R}$. A subset F of X is called *closed* if its complement $X \setminus F$ in X is open. Let $A \subset X$ and $x \in A$. We call x an *isolated point* of A if there is an open interval I containing x such that $A \cap I = \{x\}$.

Example 2.5.6 The set of all real numbers $\mathbb{R}$ with the usual order is order dense, has no least element, no last element, is separable and satisfies lub axiom.

Cantor's proof of uncountability of $\mathbb{R}$ actually proves the following. We do not repeat the same argument and invite the reader to complete the proof.

2.5 Linearly Ordered Sets

Proposition 2.5.7 *Every order dense, linearly ordered set containing more than one point and satisfying the least upper bound axiom is uncountable.*

Remark 2.5.8 This proof of uncountability of $\mathbb{R}$ due to Cantor is a proof from **THE BOOK** which the book missed.

Can every non-empty set X be linearly ordered? Using ZL we show that the answer is in the affirmative.

Lemma 2.5.9 *Let $(\mathbb{P}, \leq)$ be a partially ordered set and $x_0 \neq y_0 \in \mathbb{P}$. Suppose neither $x_0 \leq y_0$ nor $y_0 \leq x_0$. Then there is a partial order $\leq' \supset \leq$ such that $x_0 \leq' y_0$.*

Proof Take $x, y \in \mathbb{P}$. Define

$$x \leq' y \Leftrightarrow \text{either } x \leq y \text{ or } (x \leq x_0 \;\&\; y_0 \leq y).$$

Clearly, $\leq \;\subset\; \leq'$ and since $\leq$ is reflexive, $x_0 \leq' y_0$. We leave the simple proof of the fact that $\leq'$ is a partial order on X as an exercise for the reader. ∎

Proposition 2.5.10 *Let $(X, \leq)$ be a partially ordered set. Then there is a linear order $\leq'$ on X containing $\leq$.*

Proof Set

$$\mathbb{P} = \{(X, R) : \leq \;\subset\; R \;\&\; (X, R) \text{ is a poset}\}.$$

Since $(X, \leq) \in \mathbb{P}$, $\mathbb{P} \neq \emptyset$. For $(X, R), (X, S) \in \mathbb{P}$, define

$$(X, R) \leq_1 (X, S) \Leftrightarrow R \subset S.$$

Then $\leq_1$ is a partial order on $\mathbb{P}$.

Let $\mathcal{C} = \{(X, R_i) : i \in I\}$ be a chain in $\mathbb{P}$. Then $R = \cup_{i \in I} R_i$ is a partial order on X containing $\leq$ and (X, R) is an upper bound of $\mathcal{C}$. Hence, by ZL, $\mathbb{P}$ has a maximal element, say $(X, \leq')$. By the last lemma, $\leq'$ is total on X proving our result. ∎

Since every set X has a trivial partial order Δ, by the last proposition, we get the following corollary.

Corollary 2.5.11 *ZL implies that every non-empty set can be linearly ordered.*

Our next result shows that $(\mathbb{Q}, \leq)$ is a universal countable linearly ordered set, where $\leq$ is the usual order on the set of all rational numbers $\mathbb{Q}$.

Proposition 2.5.12 *Let $(X, \leq')$ be a countable linearly ordered set. Then there is an embedding $f : (X, \leq') \to (\mathbb{Q}, \leq)$.*

Proof Fix an enumeration $r_0, r_1, r_2, \ldots$ of all rational numbers where $i \neq j \Rightarrow r_i \neq r_j$. Similarly, fix an enumeration $x_0, x_1, x_2, \ldots$ of elements of X where $i \neq j \Rightarrow x_i \neq x_j$. If X is a finite set, this sequence will be finite.

We shall define $f(x_0), f(x_1), f(x_2), \ldots \in \mathbb{Q}$ by induction. Take $f(x_0) = r_0$. If $x_0 <' x_1$, choose the least i such that $r_i > r_0$ and set $f(x_1) = r_i$. Otherwise, choose the least j such that $r_j < r_0$ and put $f(x_1) = r_j$. Suppose $f(x_0), \ldots, f(x_n) \in \mathbb{Q}$ have been defined. If $x_{n+1} <' x_i$ for all $i \leq n$, choose first k such that $r_k < f(x_i)$ for all $i \leq n$ and define $f(x_{n+1}) = r_k$. If $x_i <' x_{n+1}$ for all $i \leq n$, choose first l such that $r_l > f(x_i)$ for all $i \leq n$ and put $f(x_{n+1}) = r_l$. If neither of these two hold, then there exist $1 \leq i \neq j \leq n$ such that $x_i <' x_j$, for no $m \leq n$, $x_i <' x_m <' x_j$ and $x_i <' x_{n+1} <' x_j$. Choose the least p such that $f(x_i) < r_p < f(x_j)$ and define $f(x_{n+1}) = r_p$. It is not hard to check that $f : (X, \leq') \to (\mathbb{Q}, \leq)$ is an embedding. ∎

Our next result characterizes the order type of $(\mathbb{Q}, \leq)$.

Theorem 2.5.13 *Let $(X, \leq_1)$ be a non-empty, countable, order dense linearly ordered set that has no least element and no last element. Then $(X, \leq_1)$ is order isomorphic to $(\mathbb{Q}, \leq)$.*

Proof Let $x_0, x_1, x_2, \ldots$ and $r_0, r_1, r_2, \ldots$ be enumerations of elements of X and $\mathbb{Q}$ respectively such that for $i \neq j$, $x_i \neq x_j$ as well as $r_i \neq r_j$. We shall define an isomorphism $f : (X, \leq_1) \to (\mathbb{Q}, \leq)$ by a technique called *back and forth method*.

Set $f(x_0) = r_0$. If $r_1 < r_0$, let i be the least integer such that $x_i < x_0$ and define $f(x_i) = r_1$. If $r_1 > r_0$, let i be the least integer such that $x_i > x_0$ and define $f(x_i) = r_1$.

Suppose k is even and we have carried out this process k times. This means that there exist distinct integers $i_1, \ldots, i_k$ and $j_1, \ldots, j_k$ such that $f(x_{i_1}) = r_{j_1}, \ldots, f(x_{i_k}) = r_{j_k}$ have been defined and f defined so far is order preserving. Let n be the least integer such that x_n is different from each of $x_{i_1}, \ldots, x_{i_k}$. Let m be the least integer different from each of $j_1, \ldots, j_k$ such that the extension of f to x_n defined by $f(x_n) = r_m$ is order preserving. By the order properties of $\mathbb{Q}$ such a r_m exists. Next let p be the least integer different from each of $m, j_1, \ldots, j_k$. Let q be the least integer different from $n, i_1, \ldots, i_k$ such that the extension of f defined so far to x_q by setting $f(x_q) = r_p$ is order preserving.

Proceeding similarly, the map $f : (X, \leq') \to (\mathbb{Q}, \leq)$ thus defined will be an order isomorphism. ∎

We now give a characterization of the order type of the usual order on $\mathbb{R}$.

Theorem 2.5.14 *Let $(X, \leq')$ be a non-empty, separable, order dense linearly ordered set satisfying the least upper bound axiom. Then $(X, \leq')$ is order isomorphic to $\mathbb{R}$ with usual order.*

Proof Let D be a countable dense subset of $\mathbb{R}$. Then D is order dense and has no least and no last element. Therefore, by the characterization of the order type of $(\mathbb{Q}, \leq)$, there is an order isomorphism $g : (D, \leq') \to (\mathbb{Q}, \leq)$.

2.5 Linearly Ordered Sets

Now take any $x \in X$. Since X has no last element and D is dense, $\{r \in D : r \leq' x\} \neq \emptyset$. Since X has no last element and D dense, there is a $s \in D$ such that $x <' s$. This implies that $\{g(r) : r \in D \ \& \ r \leq' x\}$ is bounded above by $g(s)$. We define $f(x) = \sup\{g(r) : r \in D \ \& \ r \leq' x\}$. We leave the rest of the proof for the reader. ∎

Remark 2.5.15 Assuming that $\mathbb{R}$ with usual order has c.c.c., we can see that $\mathbb{R}$ is separable. Fix a positive integer n. Let $\mathcal{I}_n$ be a maximal family of pairwise disjoint non-empty open intervals in $\mathbb{R}$ of length $\frac{1}{2n}$. Such a family of intervals exists by ZL. By c.c.c., $\mathcal{I}_n$ is countable. By AC, there is a countable subset D_n of $\mathbb{R}$ such that for every $I \in \mathcal{I}_n$, $D_n \cap I$ contains exactly one point. Now take any $x \in \mathbb{R}$. By the maximality of $\mathcal{I}_n$, there is a $I \in \mathcal{I}_n$ such that $(x - \frac{1}{4n}, x + \frac{1}{4n}) \cap I \neq \emptyset$. Hence, there is a $y \in D_n$ such that $|x - y| < \frac{1}{n}$. Let $D = \cup_n D_n$. Then for every $\epsilon > 0$, every open interval of length ϵ has a point in D. Since D is countable, it follows that $\mathbb{R}$ is separable.

Remark 2.5.16 Let $(X, \leq)$ be a loset which satisfies c.c.c. Then is X separable? An affirmative answer would give a better characterization of the order type of $\mathbb{R}$: *A linearly ordered set X is order isomorphic to $\mathbb{R}$ if and only if X is order dense, has no first, no last element and satisfies c.c.c. and the lub axiom.* This question was raised by the Russian mathematician Suslin. A linearly ordered set X which is order dense, has no first, no last element and satisfies c.c.c. and the lub axiom but is not separable, is called a *Suslin line*. Results on the existence of Suslin line are beyond the scope of this note. Interested readers are referred ([2, 3]) for further study.

We close this section by giving a beautiful application of the characterization of the order type of $\mathbb{Q}$.

Theorem 2.5.17 *Let C_1 and C_2 be two non-empty, closed and bounded subsets of $\mathbb{R}$ not containing any non-empty open interval and having no isolated points. Then there is an order isomorphism $f : \mathbb{R} \to \mathbb{R}$ such that $f(C_1) = C_2$.*

Proof Let $\mathcal{I}_1$ and $\mathcal{I}_2$ be the families of pairwise disjoint non-empty open intervals of $\mathbb{R}$ such that $\mathbb{R} \setminus C_i = \cup \mathcal{I}_i$, $i = 1, 2$. Since $\mathbb{R}$ is separable, both $\mathcal{I}_1$ and $\mathcal{I}_2$ are countable.

Let $a_i = \inf C_i$ and $b_i = \sup C_i$, $i = 1, 2$. If possible, suppose $a_1 \notin C_1$. Since C_1 is closed, there is an open interval (a, b) containing a_1 and disjoint from C_1. But this contradicts that $a_1 = \inf C_1$. Similarly, we see that $b_1 \in C_1$ and $a_2, b_2 \in C_2$. This implies that $(-\infty, a_i), (b_i, \infty) \in \mathcal{I}_i$, $i = 1, 2$. Set $\mathcal{J}_i = \mathcal{I}_i \setminus \{(-\infty, a_i), (b_i, \infty)\}$, $i = 1, 2$. Note that every interval in $\mathcal{J}_1$ and $\mathcal{J}_2$ is bounded.

Fix $i = 1, 2$. For interval $I, J \in \mathcal{J}_i$, define $I <_i J$ if the interval I is to the left of J. This is clearly a strict linear order on $\mathcal{J}_i$. Let $I = (a, b) <_i (c, d) = J \in \mathcal{J}_i$. If there is no open interval $I' \in \mathcal{J}_i$ that lies between I and J, then $(b, c) \subset C_i$ contradicting our hypothesis. If possible, suppose (a, b) is the least element of $\mathcal{J}_i$. Then $(-\infty, a)$ is a non-empty open interval such that $(-\infty, a) \cap C_i = \{a_i\}$. But then a_i is an isolated point of C_i which contradicts our hypothesis. Similarly, we show that $\mathcal{J}_i$ has no last element.

By the last theorem, there is an order isomorphism $H : (\mathcal{J}_1, \leq_1) \to (\mathcal{J}_2, \leq_2)$. For each $I \in \mathcal{J}_1$, fix an increasing function f_I from I onto $H(I)$. Also, fix increasing functions $f_{-\infty}$ from $(-\infty, a_1)$ onto $(-\infty, a_2)$ and f_∞ from (b_1, ∞) onto (b_2, ∞).

Set $f : \mathbb{R} \setminus C_1 \to \mathbb{R} \setminus C_2$ by

$$f = f_{-\infty} \cup \bigcup_{I \in \mathcal{J}_1} f_I \cup f_\infty.$$

Then $f : \mathbb{R} \setminus C_1 \to \mathbb{R} \setminus C_2$ is an order isomorphism.

Since C_1 and C_2 contain no non-empty open interval, $\mathbb{R} \setminus C_1$ and $\mathbb{R} \setminus C_2$ are order dense. For $x \in C_1$, we define

$$f(x) = \sup\{f(y) : y < x \ \& \ y \in \mathbb{R} \setminus C_1\}.$$

Since $\mathbb{R}$ satisfies the lub axiom, above supremum exists. It is not hard to prove that $f : \mathbb{R} \to \mathbb{R}$ is an isomorphism such that $f(C_1) = C_2$. ∎

Remark 2.5.18 Since an order isomorphism from $\mathbb{R}$ to $\mathbb{R}$ is a homeomorphism when $\mathbb{R}$ is equipped with the usual topology, f obtained in the last theorem is a homeomorphism.

2.6 Some Historical Remarks

In this section, we make some historical remarks to show how and why set theory was discovered by Cantor. This will also give the motivation for one of the most important concepts in set theory which we shall study in detail in the next section. The reader is referred to the article [4]. Most of the materials in this section appear in that article.

Cantor led to the discovery of set theory by attacking a precise problem in the then-emerging area of trigonometric series initiated by Joseph Fourier at the dawn of the nineteenth century.

Consider the formal trigonometric series

$$S \sim \sum_{n=-\infty}^{\infty} c_n e^{inx}, \quad x \in \mathbb{R}. \tag{$*$}$$

We say that the series S converges for a real number x if

$$\lim_{N \to \infty} \sum_{n=-N}^{N} c_n e^{inx}$$

exists.

2.6 Some Historical Remarks

An open problem at that time was: *If the series converges to 0 for every real number x, must then each c_n be 0?*

This problem was attempted by the great minds of that time including Dirichlet, Riemann, Lipschitz, and Heine. However, they could give an affirmative answer only for some specific cases and that too under certain conditions. It was Cantor who answered the above question in the affirmative in complete generality. Thus, Cantor proved the following result.

Theorem 2.6.1 (Cantor) *Every function $f : \mathbb{R} \to \mathbb{R}$ can have at most one representation by a trigonometric series.*

Cantor did not stop here. He could see that the hypothesis of the convergence of the trigonometric series could be relaxed on some exceptional sets of real numbers. To state Cantor's theorems precisely, we give a definition first. We call a set D of reals *a set of uniqueness* if whenever the series in $(*)$ converges to 0 for each real $x \in \mathbb{R} \setminus D$, each c_n equals 0. By Cantor's theorem, the empty set is a set of uniqueness.

Cantor gave his examples of the sets of uniqueness in terms of the derived set of a set of reals. Let $A \subset \mathbb{R}$. A real number x is called an *accumulation point* of A if for every open interval U containing x, $(A \setminus \{x\}) \cap U \neq \emptyset$. Note that if A is a closed set, then it contains all its accumulation points. If $x \in A$ is not an accumulation point of A, then it is an isolated point of A. We set

$$A' = \{x \in \mathbb{R} : x \text{ is an accumulation point of } A\}.$$

A' is called the *derived set* or the *derivative* of A. To fix ideas, *from now on, unless otherwise stated, we assume that A is a closed set of reals.* By induction, for each $n \in \mathbb{N}$, we define

$$A^{(0)} = A$$

and for every natural number n,

$$A^{(n+1)} = (A^{(n)})'.$$

Cantor proved the following result.

Proposition 2.6.2 (Cantor) *Let A be a closed subset of $\mathbb{R}$ such that $A^{(n)} = \emptyset$ for some natural number. Then A is a set of uniqueness.*

For each natural number n, Cantor could give an example of a closed set A of reals such that $A^{(n)} \neq \emptyset$ and $A^{(n+1)} = \emptyset$. He went well beyond that which we shall describe soon and which led to the discovery of set theory. We first give some very illuminating examples.

1. Let $A_1 = \{1\}$. Then $A_1 \neq \emptyset$ but $A'_1 = \emptyset$.
2. Let $A_2 = \{1 - \frac{1}{n} : n \geq 1\} \cup \{1\}$. Then $A'_2 = A_1 \neq \emptyset$ but $A''_2 = A'_1 = \emptyset$.

3. Let A_3 be obtained from A_2 by adding to it for each $n \geq 1$ an increasing sequence $\{x_k^n\}$ of real numbers all greater than $1 - \frac{1}{n}$ and $\lim_{k \to \infty} x_k^n = 1 - \frac{1}{n+1}$. Then $A_3^{(2)} = A_2^{(1)} \neq \emptyset$ and $A_3^{(3)} = (A_2^{(1)})' = \emptyset$.

4. Suppose for some $m > 2$, A_m has been defined such that $A_m^{(m-1)} \neq \emptyset$ and $A_m^{(m)} = \emptyset$. We define A_{m+1} from A_m as follows: Let $z_{n+1} > z_n$ be two consecutive isolated points of A_m. For each such pair add to A_m an increasing sequence $\{x_k^n\}$ of real numbers all greater than z_n and $\lim_{k \to \infty} x_k^n = z_{n+1}$. Call the set so obtained A_{m+1}. Then $A_{m+1}^{(m)} = A_m^{(m-1)} \neq \emptyset$ and $A_{m+1}^{(m+1)} = (A_m^{(m-1)})' = \emptyset$.

Thus, we see that for each n, there exists a compact set $A_n \subset [0, 1]$ such that $A_n^{(n)} \neq \emptyset$ and $A_n^{(n+1)} = \emptyset$. In addition, we have $A'_n = A_{n-1}$ for all $n > 0$. It is easy to see that given any closed and bounded interval $[a, b]$ and any natural number n, there exists a compact set $A_n \subset [a, b]$ such that $A_n^{(n)} \neq \emptyset$ and $A_n^{(n+1)} = \emptyset$.

Cantor continued. For each $n \geq 1$, let $A_n \subset [1 - \frac{1}{n}, 1 - \frac{1}{n+1}]$ be such that $A_n^{(n)} \neq \emptyset$ and $A_n^{(n+1)} = \emptyset$. Consider $A = \cup_{n \geq 1} A_n \cup \{1\}$. Then $A \subset [0, 1]$ is a compact set such that for every $n \geq 1$, $A^{(n)} \neq \emptyset$. Further, $A^{(n)} \supset A^{(n+1)}$. Therefore, $\cap_n A^{(n)} \neq \emptyset$ (in fact, equals $\{1\}$), is compact and its derived set is empty. Probably, for want of any better notation, Cantor set

$$A^{(\infty)} = \cap_n A^{(n)} \quad \& \quad A^{(\infty+1)} = (\cap_n A^{(n)})'.$$

Such a set A can be obtained as a subset of any closed and bounded interval $[a, b]$.

Next for any $n \geq 1$, let $A_n \subset [1 - \frac{1}{n}, 1 - \frac{1}{n+1}]$ be a copy of A obtained in the last paragraph and $B = \cup_{n \geq 1} A_n \cup \{1\}$. Then $B^{(\infty+1)} \neq \emptyset$ and $B^{(\infty+2)} = (B^{(\infty+1)})' = \emptyset$. We can carry on this process and for each $n \geq 1$, get $A_n \subset [1 - \frac{1}{n}, 1 - \frac{1}{n+1}]$ such that $A_n^{(\infty+n)} \neq \emptyset$ and $A_n^{(\infty+n+1)} = \emptyset$. Then for $A = \cup_{n \geq 1} A_n \cup \{1\}$, $A^{(\infty+\infty)} = \cap_n A^{(\infty+n)} \neq \emptyset$ and $A^{(\infty+\infty+1)} = (A^{(\infty+\infty)})' = \emptyset$.

Thus, Cantor saw that one can go on taking derivatives of a set of reals well beyond natural numbers. Although he never stated it, it is likely that he saw that the sets whose derivatives ultimately vanish are sets of uniqueness. Cantor was now faced with the task of making all these precise.

(1) Till then progress in mathematics was often being achieved at the expense of rigor. Sometimes mathematicians were unable to formulate precise definitions and appealed to intuition or geometric pictures. For instance, the only treatment of real numbers was a geometric one as "points on a line." To make his ideas precise, Cantor needed to make the set of all real numbers precise. He exchanged several letters with Dedekind and Weierstrass and finally defined the set of all real numbers as the completion of the set of all rational numbers with the usual metric. He further proved that the set of all real numbers is a complete metric space and an Archimedean ordered field. His method turned out to be very useful. For instance, for any prime number p, the field of p-adic real numbers is defined to be the completion of $\mathbb{Q}$ with respect to the p-adic norm $||\cdot||_p$.

(2) Next he was faced with the daunting task of making precise the process of *"going on taking derived sets of a set of real numbers beyond all finite stages."*

The set of all natural numbers has the properties, relevant to induction, that it is a linearly ordered set with a least element 0, every non-empty set of natural numbers has a least element and every $n > 0$ has a unique immediate predecessor. In order to extend the method of induction beyond natural numbers, the property that all natural number greater than 0 has an immediate predecessor had to go. This led Cantor to define a well-ordered set as a non-empty linearly ordered set W such that every non-empty subset of W has a least element. Cantor developed the theory of well-ordered sets and extended the method of induction to well-ordered sets. This also led to the introduction of ordinal numbers. In the next section, we shall study well-ordered sets in detail. Ordinal numbers are studied in the next chapter.

(3) Somehow Cantor felt that sets of reals whose derived sets ultimately vanish are in some sense small. He then introduced the concept of denumerable sets (now better known as countable sets) and showed that all sets whose derived sets ultimately vanish, even though they may be infinite, are countable.

(4) Cantor then asked, "Is there any set of reals which is not countable?" He then showed that, "The set of all real numbers is not denumerable." This showed him that infinite is not one—there are infinites of different sizes. After this realization, he developed the theory of cardinal numbers. In a disguised form, we have already begun the work for developing the theory of cardinal numbers. In the next chapter, we shall study cardinal numbers.

2.7 Well-Ordered Sets

A *well-ordered set* is a linearly ordered set $(W, \leq)$ such that every non-empty subset A of W contains a least element. A set X is called *well orderable* if there is a well order on X. In this case, we also say that X can be well ordered.

If $(W, \leq)$ is a well-ordered set, then W contains no infinite strictly descending sequence $x_0 > x_1 > x_2 > \ldots$. For otherwise, $A = \{x_n : n \in \mathbb{N}\}$ is a non-empty subset of W not containing a least element. Using AC, we show that the converse is also true.

Proposition 2.7.1 *Let $(W, \leq)$ be a linearly ordered set such that $\leq$ is not a well order. Then W contains a strictly decreasing sequence $x_0 > x_1 > x_2 > \ldots$.*

Proof Since $\leq$ is not a well order, W contains a non-empty subset A which contains no least element. Start by choosing any $x_0 \in A$. Since x_0 is not the least element, there is an element $x \in A$ such that $x_0 > x$. Choose any element x_1 of A such that $x_0 > x_1$. Suppose $n > 1$ assume that $x_0 > x_1 > \cdots > x_{n-1}$ in A have been chosen. Since x_{n-1} is not the least element of A, choose any $x_n < x_{n-1}$ in A. Our proof is complete by induction. ∎

Let $(W, \leq)$ be an infinite well-ordered set. Then either W has no last element (e.g., $\omega_0 = \{0, 1, 2, \ldots\}$) (when we stop) or W has a last element, say w_0. Consider

$W \setminus \{w_0\} = W(w_0)$, the initial segment of W given by w_0. If $W(w_0)$ has no last element, we stop. In this case W is a well-ordered set with no last element followed by just one element w_0.

If the initial segment $W(w_0)$ has a last element, say w_1, then W is $W(w_1)$ followed by two elements $\{w_1, w_0\}$ with $w_1 < w_0$. If $W(w_1)$ has no last element, we stop.

Proceeding similarly we will stop in a finite number of steps. Otherwise W will have a strictly decreasing sequence $\ldots < w_2 < w_1 < w_0$ which is not possible.

We have proved

Proposition 2.7.2 *A well-ordered set W is either finite or a well-ordered set with no last element followed by n many elements, where n may be 0 in case W itself has no last element.*

Example 2.7.3 $\emptyset$ (with empty order) is a well-ordered set. Since empty set has no non-empty subset, that every non-empty subset contains a least element is satisfied vacuously.

Example 2.7.4 Every linear order on a finite set is a well order. Without using AC, it can be easily proved that every finite set can be well ordered.

Example 2.7.5 The set of natural numbers $\mathbb{N}$ with the usual order is a well-ordered set.

Example 2.7.6 The set of all real numbers $\mathbb{R}$ with usual order is not a well-ordered set because the set of all positive real numbers is non-empty and does not contain a least element.

Example 2.7.7 The set

$$\{1 - 1/2, 1 - 1/3, 1 - 1/4, \ldots\} \cup \{0, 1, 2, \ldots, k-1\},$$

k a positive integer, with the usual order on $\mathbb{R}$ is a well-ordered set.

Example 2.7.8 The set

$$\{-1, -1/2, -1/3, -1/4, \ldots\} \cup \mathbb{N},$$

with the usual order on $\mathbb{R}$ is a well-ordered set.

Exercise 2.7.9 In the last section, for each $m > 0$, we defined a closed set $A_m \subset [0, 1]$ such that $A_m^{(m-1)} \neq \emptyset$ but $A_m^{(m)} = \emptyset$. Show that each of these A_m is a well-ordered set with respect to the usual order on $\mathbb{R}$.

Example 2.7.10 Let $(W_1, \leq_1)$ and $(W_2, \leq_2)$ be two well-ordered sets such that $W_1 \cap W_2 = \emptyset$. Set $W = W_1 \cup W_2$. For $x \neq y \in W$, define $x < y$ if one of the following three conditions is satisfied:

2.7 Well-Ordered Sets

1. $x, y \in W_1$ and $x <_1 y$.
2. $x, y \in W_2$ and $x <_2 y$.
3. $x \in W_1$ and $y \in W_2$.

Then $\leq$ is a well order on W. The well-ordered set W is generally denoted by $W_1 + W_2$.

Example 2.7.11 Let $(I, \leq')$ be a well-ordered set and $\{(W_i, \leq_i) : i \in I\}$ a family of well-ordered sets such that for $i \neq j \in I$, $W_i \cap W_j = \emptyset$. Take $W = \cup_{i \in I} W_i$. For $x \neq y$ in W we define $x < y$ by

1. There exists a $i \in I$ such that $x, y \in W_i$ and $x <_i y$.
2. There exist $i <' j$ in I such that $x \in W_i$ and $y \in W_j$.

Then $\leq$ is a well order on W. The well-ordered set W is generally denoted by $\sum_{i \in I} W_i$

Example 2.7.12 Let $(W_1, \leq_1)$ and $(W_2, \leq_2)$ be two well-ordered sets and $W = W_1 \times W_2$. Let $\leq$ be the lexicographic order on W, i.e., for $(a, b), (c, d) \in W$,

$$(a, b) \leq (c, d) \Leftrightarrow \text{ either } a <_1 c \text{ or } (a = c \,\&\, b \leq_2 d).$$

Then $(W, \leq)$ is a well-ordered set. In this case we denote W by $W_1 \cdot W_2$.

Let $(W, \leq)$ be a well-ordered set and $u \in W$. The set $W(u) = \{v \in W : v < u\}$ is called the *initial segment* of W given by u. Note that if u is the least element of W, then $W(u) = \emptyset$. Now let $(W_1, \leq_1)$ and $(W_2, \leq_2)$ be well-ordered sets and $f : W_1 \to W_2$ a map. If f is one-to-one and order preserving, we call f an *embedding*. If f is a bijection and order preserving, we call f an *order isomorphism* and W_1, W_2 *order isomorphic*. We shall write $W_1 \sim W_2$ if W_1 and W_2 are isomorphic and $W_1 \nsim W_2$ if they are not.

Exercise 2.7.13 Let $(W, \leq)$ be a well-ordered set. Show the following.

1. $(\{W(u) : u \in W\}, \subset)$ is a well-ordered set.
2. The map $u \to W(u)$ from $(W, \leq)$ to $(\{W(u) : u \in W\}, \subset)$ is an order isomorphism.

If $\{W_i : i \in I\}$ is a family of well-ordered sets, then we can always find a family $\{W_i' : i \in I\}$ of pairwise disjoint well-ordered sets such that for each $i \in I$, $W_i \sim W_i'$. If further I is well ordered, we define $\sum_{i \in I} W_i$ to be $\sum_{i \in I} W_i'$. Note that $\sum_{i \in I} W_i$ is well-defined modulo isomorphism. If W_1 and W_2 are well-ordered sets, $W_1 + W_2$, $W_1 \cdot W_2$ are similarly defined.

Exercise 2.7.14 Let W be a finite, no-empty well-ordered set. Show the following.

1. $W + \mathbb{N} \sim \mathbb{N}$.
2. $\mathbb{N} + W \nsim \mathbb{N}$.

It follows that if W_1 and W_2 are well-ordered sets, then $W_1 + W_2$ need not be isomorphic to $W_2 + W_1$.

Exercise 2.7.15 Let W_1, W_2 and W_3 are well-ordered sets. Show the following.

1. $W_1 \cdot W_2 \sim \sum_{u \in W_1} W_2$.
2. Show by examples that $W_1 \cdot W_2$ need not be isomorphic to $W_2 \cdot W_1$.
3. Show that $W_1 \cdot (W_2 + W_3) \sim W_1 \cdot W_2 + W_1 \cdot W_3$.

Theorem 2.7.16 *Let $(W, \leq)$ be a well-ordered set and $u \in W$. Then there is no order isomorphism $f : W \to W(u)$.*

Proof If possible, suppose there exists an order isomorphism $f : W \to W(u)$. Set $u_0 = u$ and for any $n \in \mathbb{N}$, let $u_{n+1} = f(u_n)$. Then $u_0 > u_1 > u_2 > \ldots$ is a strictly descending sequence in W. This contradicts that W is a well-ordered set and our result is proved. ∎

Theorem 2.7.17 *There does not exist a sequence of well-ordered sets $\{(W_n, \leq_n)\}$ and for each n, an element $v_n \in W_n$ such that for every $n \in \mathbb{N}$, there is an isomorphism $f_n : W_{n+1} \to W_n(v_n)$.*

Proof Suppose the above is not true and $\{f_n\}$ and $\{v_n\}$ satisfy above conditions. Set $u_0 = v_0$ and for any $n \in \mathbb{N}$, $u_{n+1} = f_0 \circ f_1 \circ \cdots f_n(v_{n+1})$. Then $\ldots <_1 u_2 <_1 u_1 <_1 u_0$ is a strictly descending sequence in W_0 contradicting that $\leq_0$ is a well order on W_0. Thus, our result is proved. ∎

We now turn our attention to extending the process of induction beyond natural numbers. As we have seen in the last section, this was the first major set theoretic result for mathematics.

We shall first recall the method of induction on natural numbers. Using axioms of ZF, we broadly defined the set of all natural numbers $\mathbb{N}$. We also asserted (though did not give any proof) that using ZF, we can show that $\mathbb{N}$ satisfies Peano axioms.

We use the method of induction on natural numbers for two reasons:

1. Given a sequence $\{\varphi_n\}$ of statements (formally in the language of set theory), by the method of induction we prove that for all n, φ_n is true.
2. By method of induction we define a sequence in a set satisfying some properties.

But the method has to be valid in ZF. Since using ZF it is proved that $\mathbb{N}$ satisfies Peano's axioms, whatever is proved for natural numbers using Peano axioms is valid in ZF also.

Proposition 2.7.18 (Proof by induction) *Let $\{\varphi_n\}$ be a sequence of statements such that φ_0 is true and for every n, $\varphi_n \to \varphi_{n+1}$ is true. Then $\forall n \varphi_n$ is true.*

Proof Let A be the set of all natural numbers such that φ_n is not true. We are required to show that $A = \emptyset$. Suppose not. Then A has a least element, say k. Since φ_0 is true, $k > 0$. Hence, $k = m + 1$ for some natural number m. Since $m \notin A$, φ_m is true. By our assumptions, $\varphi_k = \varphi_{m+1}$ is true. This contradiction proves our result. ∎

Formally speaking we have proved the following result in ZF.

2.7 Well-Ordered Sets

Proposition 2.7.19 *For every sequence $\{\varphi_n\}$ of statements, the following holds:*

$$ZF \vdash (\varphi_0 \wedge \forall n(\varphi_n \to \varphi_{n+1})) \to \forall n \varphi_n.$$

However, we shall not be so formal later.

Proposition 2.7.20 (Definition by induction) *Let X be a set, $a \in X$ and $g : X \to X$ any function. Then there is a unique sequence $\{x_n\}$ in X such that $x_0 = a$ and for every $n \in \mathbb{N}$, $x_{n+1} = g(x_n)$.*

Proof For each $n \in \mathbb{N}$, let φ_n be the statement, "there is a unique function $h_n : \{0, 1, \ldots, n\} \to X$ such $h_n(0) = a$ and for all $i < n$, $h_n(i+1) = g(h_n(i))$."

φ_0 is true because $h_0(0) = a$ is the only function witnessing that φ_0 is true.

Take any natural number n and assume that φ_n is true. So we have a unique function $h_n : \{0, 1, \ldots, n\} \to X$ witnessing φ_n. Let $h_{n+1} : \{0, 1, \ldots, n+1\} \to X$ be the extension of h_n such that $h_{n+1}(n+1) = g(h_n(n))$. Then $h_{(n+1)}(0) = a$ and for every $i < n+1$, $h_{n+1}(i+1) = g(h_{n+1}(i))$. If $h'_{n+1} : \{0, 1, \ldots, n+1\} \to X$ is another such function, then

$$h'_{n+1}|\{0, 1, \ldots, n\} = h_n = h_{n+1}|\{0, 1, \ldots, n\}$$

by the uniqueness of h_n and $h'_{n+1}(n+1) = g(h_n(n)) = h_{n+1}(n+1)$. Thus we see that φ_{n+1} is true.

Hence, by Proposition 2.7.18, for every n, φ_n is true. Now set $x_n = h_n(n), n \in \mathbb{N}$. This sequence $\{x_n\}$ is the unique sequence with desired properties. ∎

$\mathbb{N}$ has a property that every $m > 0$ has an immediate predecessor. This is not satisfied by all well-ordered sets. For instance, if

$$W = \{-1, -1/2, -1/3, \ldots\} \cup \{0\}$$

with the usual order, then W is a well-ordered set, 0 is not the first element of W and has no immediate predecessor. Therefore, for induction on general well-ordered sets, one modifies the method of induction to what is known as the method of complete induction in case of $\mathbb{N}$.

Theorem 2.7.21 (Proof by transfinite induction) *Let $\{\varphi_v : v \in W\}$ be a family of sentences indexed by W. Suppose for every $u \in W$, $(\forall v < u \varphi_v) \to \varphi_u$ holds. Then $\forall u \varphi_u$ holds.*

Proof Let $A = \{v \in W : \varphi_v \text{ is not true}\}$. If possible, Suppose $A \neq \emptyset$. Since W is well ordered, let u be the least element of A. Then for all $v < u$, φ_v holds. Hence, by our assumption, φ_u holds. This is a contradiction. ∎

Let $(W, \leq)$ be a well-ordered set and X a non-empty set. We set $\mathcal{F}$ to be the set of all functions with domain an initial segment of W and the range contained in X.

Theorem 2.7.22 (Definition by transfinite induction) *Let $(W, \leq)$, X and $\mathcal{F}$ be as above. Given any function $G : \mathcal{F} \to X$, there is a unique function $f : W \to X$ such that for all $u \in W$, $f(u) = G(f|W(u))$.*

Proof For each $v \in W$, let φ_v be the statement, "there is a unique function $f_v : \{w \in W : w \leq v\} \to X$ such that for all $w \leq v (f_v(w) = G(f_v|W(w)))$." Let $u \in W$ and φ_v hold for all $v < u$. By uniqueness, we can see that for $v < v' < u$, $f_{v'}|\{w \in W : w \leq v\} = f_v$. Hence, $g = \cup_{v<u} f_v$ is a well-defined function from $W(u)$ to X. Take f_u to be the extension of g such $f_u(u) = G(g)$. It is now easy to argue that φ_u holds.

Therefore by Theorem 2.7.21, for every $u \in W$, φ_u holds. Let $\{f_u : u \in W\}$ be the set of functions so obtained. Now take $f = \cup_{u \in W} f_u$ to complete the proof. ∎

Exercise 2.7.23 Let $(W, \leq)$ be a well-ordered set and $f : W \to W$ a function such for all $x, y \in W$, $x < y \Rightarrow f(x) < f(y)$. Show that for all $x \in W$, $x \leq f(x)$.

Let W_1 and W_2 be two well-ordered sets. Define $W_1 \prec W_2$ if W_1 is order isomorphic to an initial segment of W_2 and write $W_1 \preceq W_2$ if either $W_1 \prec W_2$ or $W_1 \sim W_2$ holds.

The trivial proof of the next proposition is omitted.

Proposition 2.7.24 *Let $(W, \leq)$ be a well-ordered set. Let $* \notin W$. Set $W' = W \cup \{*\}$. Define $<^*$ on W' by $<^* |W = <$ and for every $x \in W$, $x <^* *$. Then $(W', <')$ is a woset and $(W, <) \prec (W', <')$.*

Proposition 2.7.25 (Trichotomy theorem) *Let W_1 and W_2 be two well-ordered sets. Then exactly one of the following three conditions holds.*

$$W_1 \prec W_2 \text{ or } W_1 \sim W_2 \text{ or } W_2 \prec W_1.$$

Proof If possible, suppose both $W_1 \sim W_2$ and $W_2 \prec W_1$ simultaneously hold. Therefore, there exists an $u \in W_1$ and there exist order isomorphisms $f : W_1 \to W_2$ and $g : W_2 \to W_1(u)$. Then $g \circ f$ is an order isomorphism from W_1 onto an initial segment of itself. We have shown earlier that this is impossible.

Next let $u \in W_1$ and $v \in W_2$. Suppose there exist order isomorphisms $f : W_1 \to W_2(v)$ and $g : W_2 \to W_1(u)$. Then $g \circ f$ is an order isomorphism from W_1 onto an initial segment of itself, which again is a contradiction.

We have now shown that at most one of the three above conditions holds. We shall now show that at least one of the above three conditions holds.

Set $X = W_2 \cup \{*\}$, where $* \notin W_2$ and $\mathcal{F}$ the set of all functions whose domain is an initial segment of W_1 and range contained in X.

Take any $g \in \mathcal{F}$. If $W_2 \setminus \text{range}(g) \neq \emptyset$, we define $G(g)$ to be the least element of $W_2 \setminus \text{range}(g)$. Otherwise, we define $G(g) = *$. By the proposition 'definition by transfinite induction' there is a unique function $f : W_1 \to X$ such that for every $w \in W_1$, $f(w) = G(f|W_1(w))$.

Consider the case when $*$ belongs to the range of f. Let $u \in W_1$ be the least element of W_1 such that $f(u) = *$. It is then easy to check that $f(W_1(u)) = W_2$ and that $f : W_1(u) \to W_2$ is an isomorphism. So in this case $W_2 \prec W_1$ holds.

2.8 Equivalence of ZL, WOP, and AC

Next consider the case when $*$ does not belong to the range of f. Then $f : W_1 \to W_2$ is an embedding with range W_2 or an initial segment of W_2. So one of the two conditions $W_1 \sim W_2$ or $W_1 \prec W_2$ holds.

Our proof is complete now. ∎

Exercise 2.7.26 Show that $W_1 \preceq W_2$ if and only there is an embedding from W_1 into W_2.

Exercise 2.7.27 Let $\{(W_n, \leq_n)\}$ be a sequence of countable well, ordered sets and $W = \sum_n W_n$. Then show that W is a countable well-ordered set such that for every $n \in \mathbb{N}$, $(W_n, \leq_n) \preceq (W, <)$. Give an example where $W \sim W_m$ for some m.

Here is a fundamental question. Can every set be well-ordered? Cantor tried very hard but failed to define a well order on the set $\mathbb{R}$ of all real numbers. Let X be a set and $(W, \leq)$ a well-ordered set. Suppose there exists an injection $f : X \to W$. For $x, y \in X$, define $x \leq' y \Leftrightarrow f(x) \leq f(y)$. Then $\leq'$ is a well order on X. From here it is easy to see that every countable set can be well ordered. Does there exist an uncountable well-ordered set? We can also ask if there exists a well-ordered set W such that there is an injection $f : \mathbb{R} \to W$?

The statement that "every set can be well ordered" is known as the *well-ordering principle*, WOP in short.

Well-ordering principle (WOP). *Every set X can be well ordered.*

In the next section, we shall prove that ZL, WOP, and AC are all equivalent (in ZF).

2.8 Equivalence of ZL, WOP, and AC

Theorem 2.8.1 *The following three statements are equivalent.*

1. ZL.
2. WOP.
3. AC.

Proof Assume ZL and take any non-empty set X. Let

$$\mathbb{P} = \{(W, \leq) : W \subset X \ \& \ \leq \text{ is a well order on } W\}.$$

Since the empty relation belongs to $\mathbb{P}$, $\mathbb{P} \neq \emptyset$. For $(W, \leq) \neq (W', \leq')$ in $\mathbb{P}$, set $(W, \leq) < (W', \leq')$ if $<' | W = <$ and $(W, \leq)$ is an initial section of $(W', \leq')$. Clearly, $(\mathbb{P}, <)$ is a strict non-empty partially ordered set.

Let $\mathcal{C} = \{(W_i, \leq_i) : i \in I\}$ be a chain in $\mathbb{P}$. For $i \neq j \in I$, either $(W_i, \leq_i)$ is an initial section of $(W_j, \leq_j)$ or $(W_j, \leq_j)$ is an initial section of $(W_i, \leq_i)$. In either case, both W_i and W_j have the same first element. Now consider $W = \cup_{i \in I} W_i$. By the definition of $<$ on $\mathbb{P}$, $\leq$ defined on W by $\leq |W_i = \leq_i$ for all $i \in I$ is well defined and a linear order on W.

Let $\emptyset \neq A \subset W$. Take an $i \in I$ such that $A \cap W_i \neq \emptyset$. Suppose a is the $\leq_i$ least element of $A \cap W_i$. Let $j \neq i \in I$ be such that $A \cap W_j \neq \emptyset$. Then either $(W_i, \leq_i)$ is an initial section of $(W_j, \leq_j)$ or $(W_j, \leq_j)$ is an initial section of $(W_i, \leq_i)$. In either case we see that the least elements of $A \cap W_i$ and $A \cap W_j$ are the same. This implies that a is the least element of A with respect to $\leq$. It is also not hard to see that each $(W_i, \leq_i)$ is either an initial segment of $(W, \leq)$ or equals $(W, \leq)$. Hence, $(W, \leq) \in \mathbb{P}$ is an upper bound of $\mathcal{C}$.

By ZL, $\mathbb{P}$ has a maximal element, say $(W_0, \leq_0)$. If possible, suppose $W_0 \neq X$. Let $x \in X \setminus W_0$. Set $W_1 = W_0 \cup \{x\}$, Define $\leq_1$ on W_1 by $\leq_1 |W_0 = \leq_0$ and for every $y \in W_0$, $y <_1 x$. Then $(W_0, \leq_0) < (W_1, \leq_1)$ contradicting the maximality of $(W_0, \leq_0)$. Hence, $\leq_0$ is a well order on X proving that ZL implies WOP.

Next assume WOP. Let $\{A_i : i \in I\}$ be a family of non-empty sets. By WOP, there is a well order $\leq$ on $\cup_{i \in I} A_i$. Define $f(i) \in A_i$ to be the least element of A_i with respect to $\leq$. Thus, we see that WOP implies AC.

It remains to prove that AC implies ZL. We assume AC. Fix a non-empty partially ordered set $(\mathbb{P}, \leq)$ such that every chain in $\mathbb{P}$ has an upper bound in $\mathbb{P}$. We are required to show that $\mathbb{P}$ has a maximal element.

Suppose we can find a chain C in $\mathbb{P}$ so that there is no $y \in \mathbb{P}$ such that for every $x \in C$, $x < y$. By our hypothesis, C has an upper bound, say x_0. Then x_0 will be a maximal element of $\mathbb{P}$ and our proof will be complete. For each chain C, define

$$C' = \{y \in \mathbb{P} : \forall x \in C (x < y)\},$$

and

$$\mathcal{C}_0 = \{C : C \text{ is a chain \& } C' \neq \emptyset\}.$$

By AC, there is a function $f : \mathcal{C}_0 \to \mathbb{P}$ such that $f(C) \in C'$ for every $C \in \mathcal{C}_0$. For each chain C in $\mathbb{P}$, we define its *successor* $s(C)$ by

$$s(C) = \begin{cases} C \cup \{f(C)\} & \text{if } C \in \mathcal{C}_0 \\ C & \text{otherwise.} \end{cases}$$

We need to show that there is a chain C such that $s(C) = C$. We call a family $\mathcal{M}$ of chains a *normal family* if the following three conditions are satisfied.

1. $\emptyset \in \mathcal{M}$.
2. Whenever a chain $C \in \mathcal{M}$, so is $s(C)$.
3. For every subfamily $\{C_i : i \in I\}$ of $\mathcal{M}$, whenever $C = \cup_{i \in I} C_i$ is a chain, $C \in \mathcal{M}$.

Clearly, the set $\mathcal{C}$ of all chains in $\mathbb{P}$ is a normal family. Further, the intersection of any set of normal families is a normal family. Let $\mathcal{N}$ be the intersection of all normal families. Then $\mathcal{N}$ is a normal family which is contained in every normal family.

Main Observation. For every $C, D \in \mathcal{N}$, either $C \subset D$ or $D \subset C$.

2.8 Equivalence of ZL, WOP, and AC

Assuming this we complete the proof first. Let $C = \cup\{D : D \in \mathcal{N}\}$. By the above observation, C is a chain. Since $\mathcal{N}$ is normal $C \in \mathcal{N}$. Further, C is the largest element of $\mathcal{N}$. Since $\mathcal{N}$ is normal, $S(C) \in \mathcal{N}$. Therefore, $s(C) \subset C \subset s(C)$ implying that $C = s(C)$, completing the proof.

We now proceed to prove the main observation. Call a chain C in $\mathcal{N}$ *good* if for every $D \in \mathcal{N}$ either $C \subset D$ or $D \subset C$. It is sufficient to prove that every chain $C \in \mathcal{N}$ is good.

The following is a crucial fact.

Fact. If C is good then for every $N \in \mathcal{N}$, either $N \subset C$ or $s(C) \subset N$.

Assuming the fact we complete the proof of the main observation now. Toward proving this, let

$$\mathcal{M}_1 = \{C \in \mathcal{N} : C \text{ is good}\}.$$

We have the following.

1. Since the empty chain $\emptyset \in \mathcal{M}_1$.
2. Let $\{C_i : i \in I\} \subset \mathcal{M}_1$ and $C = \cup_{i \in I} C_i$ be a chain. Take any $D \in \mathcal{N}$. If each $C_i \subset D$, then $C \subset D$. Now consider the case that some $C_i \not\subset D$ for some $i \in I$. Since C_i is good, it follows that $D \subset C_i \subset C$.
3. Now let $C \in \mathcal{M}_1$, i.e., C is good and $D \in \mathcal{N}$. Then by the above fact either $D \subset C \subset s(C)$ or $s(C) \subset D$. Thus, $s(C) \in \mathcal{M}_1$.

The above arguments show that $\mathcal{M}_1 \subset \mathcal{N}$. Since $\mathcal{N}$ is the smallest normal family, $\mathcal{M}_1 = \mathcal{N}$. This proves the main observation.

It remains to prove the fact now.

Let C be a good chain. Consider

$$\mathcal{M}_2 = \{N \subset \mathcal{N} : N \subset C \text{ or } s(C) \subset N\}.$$

Then

1. Clearly the empty chain $\emptyset \in \mathcal{M}_2$.
2. Let $\{N_i : i \in I\} \subset \mathcal{M}_2$ and $N = \cup_{i \in I} N_i$ be a chain. Then $N \in \mathcal{N}$ because $\mathcal{N}$ is normal. If each $N_i \subset C$, $N \subset C$. Otherwise some $N_i \not\subset C$. Since $N_i \in \mathcal{M}_2$, $s(C) \subset N_i \subset N$. Hence $N \in \mathcal{M}_2$.
3. Now let $N \in \mathcal{M}_2$. We prove that $s(N) \in \mathcal{M}_2$. If $s(C) \subset N$, $s(C) \subset s(N)$. So assume that $N \subset C$. Since C is good and $s(N) \in \mathcal{N}$, either $s(N) \subset C$ or $C \subset s(N)$. In the first case, $s(N) \in \mathcal{M}_2$. In the second case, we have $N \subset C \subset s(N)$. Since $s(N)$ differs from N by at most one point, either $C = N \subset s(N)$ or $C = s(N)$ implying $s(N) \in \mathcal{M}_2$.

We have now proved that $\mathcal{M}_2$ is normal implying that $\mathcal{M}_2 = \mathcal{N}$. Thus, we have proved the fact also, completing the proof of $AC \Rightarrow ZL$. ∎

References

1. S.M. Srivastava, *A Course on Borel Sets, GTM 180* (Springer, New York, 1997)
2. T. Jech, *Set Theory*, Springer Monographs in Mathematics, 3rd edn. (Springer, New York, 2002)
3. K. Kunen, *Set Theory: An Introduction to Independence Proofs* (North-Holland Publishing Company, 1980)
4. S.M. Srivastava, *How did Cantor Discover Set Theory and Topology, Resonance*, vol. 19, No. 11, 977-999. Indian Academy of Sciences and Springer (2014)

Chapter 3
More on Cardinals and Ordinals

By now we have made sufficient progress to introduce ordinal and cardinal numbers, the two most fundamental concepts in set theory. In this chapter, we shall make a detailed study of ordinal and cardinal numbers.

3.1 Ordinal Numbers—An Informal Introduction

Let ON be a class of sets such that ON consists of well-ordered sets satisfying the following two properties:

(i) Well-ordered sets in ON are pairwise non-isomorphic.
(ii) Every well-ordered set is order isomorphic to a well-ordered set in ON.

In the next section, we shall define such a class precisely. For the time being, we assume that such a class exists and continue with our study.

Since no two well-ordered sets in ON are order isomorphic, this implies that every well-ordered set is order isomorphic to exactly one well-ordered set in ON. Well-ordered sets belonging to ON are called *ordinal numbers*. Well-ordered sets belonging to ON, i.e., ordinal numbers, will be denoted by α, β, γ with or without suffix. By the properties of well-ordered sets proved in Chap. 2, given any two distinct $\alpha, \beta \in ON$, either α is order isomorphic to an initial segment of β or β is order isomorphic to an initial segment of α.

We set $\alpha < \beta$ if α is order isomorphic to an initial segment of β. We shall also write $\alpha \leq \beta$ if either $\alpha < \beta$ or $\alpha = \beta$.

By the general properties of well-ordered sets, it follows that $\leq$ is reflexive, antisymmetric, transitive, and total. Also, by Theorem 2.7.17, there is no strictly descending infinite sequence $\cdots < \alpha_2 < \alpha_1 < \alpha_0$ of ordinal numbers. Hence, any set of ordinal numbers is a well-ordered set.

ON has a least element. Otherwise, we can choose a descending infinite sequence of ordinals $\cdots < \alpha_2 < \alpha_1 < \alpha_0$. (Here we are using AC. However, in the precise definition of ON that will not be the case.) The smallest ordinal number is denoted by 0. Note that 0 must be the empty set. For if $x \in 0$, the ordinal number isomorphic to the initial segment $0(x)$ is smaller than 0, a contradiction.

Let $(W, \leq)$ be a well-ordered set and $W' = W \cup \{\infty\}$ with $\infty \notin W$. We define a binary relation $<'$ on W' by $<' |W = <$ and $x <' \infty$ for all $x \in W$. Then $\leq'$ is a well order on W' with $W = W'(\infty)$, the initial segment of W' defined by ∞. By Theorem 2.7.16, there is no largest ordinal number.

Hence, there is a smallest ordinal number, to be denoted by 1, greater than 0. We can argue as before and see that 1 consists of exactly one point. We can proceed inductively and define "finite" ordinals $0, 1, 2, \ldots$.

We see that $(\{0, 1, 2, \ldots\}, \leq)$ is a well-ordered set. Hence, there is an ordinal number, denoted by ω_0, order isomorphic to $(\{0, 1, 2, \ldots\}, \leq)$. Note that ω_0 has no last element but each of $1, 2, 3, \ldots$ does.

An ordinal number $\alpha \neq 0$ which has no last element is called a *limit ordinal*, otherwise it is called a *successor ordinal*. Let β be a successor ordinal and x its last element. Then there is a unique ordinal number, say α, order isomorphic to $\beta(x)$. We then write $\beta = \alpha + 1$.

In Proposition 2.7.2, we saw that every well-ordered set is a well-ordered set with no last element followed by n many elements, where n may be 0. Thus, every ordinal number β is order isomorphic to a finite well-ordered set or isomorphic to a limit ordinal number, say α, followed by n many elements. Hence, every ordinal number β is of the form $\alpha + n$, where α is either 0 or a limit ordinal. We call β an *even ordinal* if n is even and an *odd ordinal* if n is odd. In particular, 0 and limit ordinals are even ordinals.

An ordinal number α is called *finite* if the underlying set of α is finite; it is called *countable* (*uncountable*) if the underlying set of α is countable (resp. uncountable).

A subtle point: Let α be an ordinal number. Then for an ordinal number β, $\beta < \alpha$ if and only if β is order isomorphic to a unique initial segment $\alpha(x)$, $x \in \alpha$, of α. Next let $\beta, \gamma < \alpha$. Let β be order isomorphic to $\alpha(x)$ and γ to $\alpha(y)$, where $x, y \in \alpha$. Let $<'$ denote the order on α. Then

$$\beta < \gamma \Leftrightarrow x <' y.$$

Thus, $(\{\beta : \beta < \alpha\}, <)$ is order isomorphic to $(\alpha, <')$.

3.2 Making Concepts Precise

The main imprecision comes from the fact that we do not know which well-ordered sets have been put in ON. Considering the discussion in the last section, few things are fairly clear:

3.2 Making Concepts Precise

1. 0 has to be the empty set (with empty order). Hence, $\emptyset \in ON$.
2. 1 has to be a singleton. So, we may as well choose $1 = \{0\} = \{\emptyset\}$.
3. 2 has to consist of two elements in some order. We may choose $2 = \{0, 1\}$ with $0 < 1$, the ordinal ordering that already exists between 0 and 1. This is consistent with the subtle observation that we made above. Note that

$$2 = \{\emptyset, \{\emptyset\}\} = 1 \cup \{1\} = \{0\} \cup \{1\} = \{0, 1\}.$$

Further, $1 = s(\emptyset)$ and $2 = s(1)$.

4. For any $n > 0$, we can take $n = s(n-1) = \{0, 1, \cdots, n-1\}$ with the ordering between $0, 1, \cdots, n-1$ that has already been defined. Note that this is also consistent with the subtle observation that we made in the end of the last section.
5. For ω_0, it is quite natural to choose $\{0, 1, 2, \ldots\}$ with the ordering between these that already exists. This is the first case of a limit ordinal. We may adopt the same policy for all limit ordinals α, namely we take $\alpha = \{\beta \in ON : \beta < \alpha\}$ and for successor ordinals $\beta = \alpha + 1$, we may take $\beta = \alpha \cup \{\alpha\} = s(\alpha)$.

Clearly, some induction is going on. But it is difficult to put this elegantly. So, we think of the class of all sets that we get this way. We try to picture the sets $0, 1, 2, \cdots, \omega_0, \omega_0 + 1$ as far as possible.

Call a set x *transitive* if whenever a set $y \in x$, $y \subset x$, i.e., $\forall z(z \in y \Rightarrow z \in x)$. For instance, consider

$$10 = \{0, 1, 2, \cdots, 9\}.$$

Take, say $7 = \{0, 1, 2, \cdots, 6\} \in 10$. All the elements of 7 are elements of 10. It is easy to see that 10 is a transitive set.

Consider $\omega_0 = \{0, 1, 2, \ldots\}$. Take any element $\{0, 1, \cdots, n-1\} = n \in \omega_0$. All the elements of n are also the elements of ω_0. Now consider

$$\omega_0 + 1 = s(\omega_0) = \{0, 1, \cdots, \{0, 1, 2, \ldots\}\}.$$

Take $\omega_0 = \{0, 1, 2, \ldots\} \in \omega_0 + 1$. All the elements of ω_0 are clearly elements of $\omega_0 + 1$. It can be checked easily that $0, 1, 2, \ldots, \omega_0, \omega_0 + 1$ are all transitive.

Next we note another fact about these sets. Take any two distinct elements $x \neq y$ in any of these sets; for example consider

$$\omega_0 + 1 = \{0, 1, 2, \cdots, \{0, 1, 2, \ldots\}\}.$$

Note that for any two distinct elements $x \neq y \in \omega_0 + 1$, $x < y \Leftrightarrow x \in y$. For instance if $y = \omega_0 \in \omega_0 + 1$ and $x \neq y$ another element of $\omega_0 + 1$, then $x < y \Leftrightarrow x \in y$. In particular, $\in$ restricted to any of these sets coincides with the well order on the set that they are already equipped with. In particular, $\in$ restricted to all these sets are strict linear orders.

By the *the foundation axiom* of ZF there is no infinite sequence of sets $x_0, x_1, x_2 \ldots$ such that $\cdots \in x_2 \in x_1 \in x_0$. Thus if $\epsilon | x$ is a strict linear order, it is already a well order.

Let ON be the collection of all sets α such that α is transitive and $\in | \alpha$ is a strict linear order.

Theorem 3.2.1 *ON is a class.*

Proof Let $tr(x)$ denote the formula

$$\forall y(y \in x \to \forall z(z \in y \to z \in x))$$

and $lo(x)$ the formula

$$\forall y \forall z((y \in x \land z \in x) \to (y \in z \lor z \in y))$$

$$\land \forall y \forall z \forall v((v \in z \land z \in y \land y \in x) \to v \in y).$$

Then ON is defined by the formula

$$tr(x) \land lo(x).$$

∎

Lemma 3.2.2 *Let $\alpha \in ON$ and $\beta \in \alpha$. Then $\beta \in ON$.*

Proof Assume that α and β are as above. Since α is transitive, $\beta \subset \alpha$. Therefore, as $\in | \alpha$ is a strict linear order, $\in | \beta$ is a strict linear order. It remains to show that β is transitive. Since α is transitive and $\beta \in \alpha$, $\beta \subset \alpha$. Take any $\delta \in \beta$. Then $\delta \in \alpha$. Now take any $\gamma \in \delta$. By the same argument $\gamma \in \alpha$. Thus, β, δ and γ are all in α. Further, $\gamma \in \delta$ and $\delta \in \beta$. Since $\in | \alpha$ is a linear order, it follows that $\gamma \in \beta$. Thus, β is transitive. ∎

Proposition 3.2.3 *Let $\alpha \neq \beta \in ON$. Then either $\alpha \in \beta$ or $\beta \in \alpha$.*

Proof We have $(\alpha, \in | \alpha)$ and $(\beta, \in | \beta)$ are well-ordered sets. If they were isomorphic, then clearly $\alpha = \beta$ which is not the case. Hence, by Proposition 2.7.25, either $(\alpha, \in | \alpha)$ is isomorphic to an initial segment of β or $(\beta, \in | \beta)$ is isomorphic to an initial segment of α. Without any loss of generality assume that $(\alpha, \in | \alpha)$ is isomorphic to the initial segment $\beta(\gamma)$, where $\gamma \in \beta$. Any such isomorphism has to be identity on α. Now it follows that $\alpha = \gamma \in \beta$. ∎

Exercise 3.2.4 Let A be a set of ordinals. Show that $\cup_{\alpha \in A} \alpha$ is an ordinal number.

3.2 Making Concepts Precise

It is more traditional to state this result as follows.

Theorem 3.2.5 (Trichotomy theorem) *For $\alpha, \beta \in ON$, exactly one of the following three properties holds.*

$$\alpha < \beta \text{ or } \alpha = \beta \text{ or } \beta < \alpha.$$

Proposition 3.2.6 (Burali-Forti Theorem) *ON is not a set.*

Proof If possible suppose ON is a set. Then ON is a transitive set such that $\in |ON$ is a strict linear order implying that $ON \in ON$. This contradicts the foundation axiom of ZF. ∎

Theorem 3.2.7 *Let $(W, \leq)$ be a well-ordered set. Then there is a unique $\alpha \in ON$ such that W and α are isomorphic.*

Proof The uniqueness part follows from the trichotomy Theorem 3.2.5.

We define a function f on W by transfinite induction. Let $u_0 \in W$ be the least element of W. We put $f(u_0) = \emptyset = 0$. Suppose $u \in W$ is not the least element and $f|W(u)$ has been defined. If u has no immediate predecessor, then we define $f(u) = \{f(v) : v < u\}$. Next suppose u has an immediate predecessor, say v. Then we define $f(u) = s(f(v))$. Note that for every $u \in W$, $f(u) = \{f(v) : v < u\}$.

By the replacement axiom, the range of f is a set, say α. We invite the reader to prove the following:

1. $\alpha \in ON$.
2. $f : (W, \leq) \to (\alpha, \in |\alpha)$ is an isomorphism. ∎

The unique ordinal α obtained in the last theorem will be called the *order type* of the well-ordered set $(W, \leq)$.

From now on, each $\alpha \in ON$ will be assumed to be equipped with the binary relation $\in$ and will be called an ordinal number. The following result is obvious.

Proposition 3.2.8 *ω_0 is the first infinite ordinal.*

Let $\alpha, \beta \in ON$. Then there is a unique ordinal $\gamma \in ON$ such that $\alpha + \beta \sim \gamma$. (Recall that in the section on well-ordered sets, given two well-ordered sets W_1 and W_2, we had defined their sum $W_1 + W_2$. So, $\alpha + \beta$ makes sense.) We denote the unique ordinal $\gamma \sim \alpha + \beta$ by $\alpha + \beta$ itself.

Using Proposition 2.7.2, we have the following result.

Theorem 3.2.9 *Every $\alpha \in ON$ has unique representation $\alpha = \beta + n$ where $\beta = 0$ or has no last element and $n \in \omega_0$.*

Exercise 3.2.10 If $0 < n$ is a finite ordinal and α an infinite ordinal, show that

$$n + \alpha = \alpha < \alpha + n.$$

An ordinal number α is called a *successor ordinal* if there is an ordinal β such that $\alpha = \beta + 1$. If $\alpha \neq 0$ and has no last element element, then α is called a *limit ordinal*. An ordinal number $\alpha = \beta + n$, $\beta = 0$ or limit and $n \in \omega_0$ is called an *even* (*odd*) ordinal if n is even (resp. odd).

Proposition 3.2.11 *Let $A \subset \mathbb{R}$ be such that $< | A$ is a well-order on A, where $<$ is the usual order order on $\mathbb{R}$. Then A is countable.*

Proof If possible, suppose A is uncountable. Without any loss of generality, let A be order isomorphic to the ordinal number α. Then

$$\{(\beta, \beta + 2) : \beta \text{ even and } \beta + 2 \in \alpha\}$$

is an uncountable family of pairwise disjoint, non-empty, open intervals in $\mathbb{R}$. This is a contradiction. ∎

Exercise 3.2.12 Let α be a limit ordinal. Show that $\alpha = \cup \alpha$.

Lemma 3.2.13 *For every countable limit ordinal α there is a strictly increasing sequence $\{\beta_n\}$ of countable ordinals such that $\alpha = \sup \beta_n$.*

Proof Since α is countable, fix an enumeration $\{\alpha_n : n \in \mathbb{N}\}$ of $\{\delta \in ON : \delta \in \alpha\}$. Set $\beta_0 = \alpha_0$. Suppose for some $n \geq 0$, $\beta_0 < \beta_1 < \cdots < \beta_n < \alpha$ have been defined. Since α is a limit ordinal, $A = \{i \in \mathbb{N} : \beta_n < \alpha_i\} \neq \emptyset$. Let i_0 be the least element of A. Set $\beta_{n+1} = \alpha_{i_0}$. The rest of the proof is easy to see. ∎

Remark 3.2.14 The last lemma uses AC.

Proposition 3.2.15 $\omega_1 = \{\alpha \in ON : \alpha \text{ is countable}\}$ *is a set.*

Proof Let $\mathcal{A}$ be the set of all $(A, \leq)$ where either $A = \{0, 1, \cdots, n-1\}$ for some $n \in \mathbb{N}$ or $A = \mathbb{N}$ and $\leq$ is a well order on A. Then $\mathcal{A}$ is a set. For $(A, \leq) \in \mathcal{A}$, set

$$[(A, \leq)] = \{(B, \leq') \in \mathcal{A} : (A, \leq) \sim (B, \leq')\}$$

and
$$\mathcal{A}/\sim = \{[(A, \leq)] : (A, \leq) \in \mathcal{A}\},$$

where $\sim$ denotes order isomorphism. Then $\mathcal{A}/\sim$ is a set. By the properties of ON, there is a unique function f on $\mathcal{A}/\sim$ such that for every $[(A, \leq)] \in \mathcal{A}/\sim$, $\alpha = f([(A, \leq)]) \in ON$ and $\alpha \sim (A, \leq)$. The range of f is a set by the replacement axiom and equals ω_1. ∎

Proposition 3.2.16 $\omega_1 = (\omega_1, \in)$ *is the least uncountable ordinal.*

Proof If ω_1 were countable, $\omega_1 \in \omega_1$. This contradiction proves that ω_1 is an uncountable ordinal. Since each $\alpha \in \omega_1$ is countable, clearly ω_1 is the least uncountable ordinal. ∎

3.3 Cardinal Numbers

We shall study ω_1 in detail later in these notes.

Since given any well-ordered set $(W, \leq)$ there is a unique ordinal number $\alpha \in ON$ isomorphic to $(W, \leq)$, it is sufficient to consider methods of transfinite induction on ordinals only. We restate methods of transfinite induction once again without any proof.

Theorem 3.2.17 (Proof by transfinite induction) *Let α be an ordinal number and $\{\varphi_\beta : \beta < \alpha\}$ be a transfinite sequence of statements of length α. Suppose for every $\beta < \alpha$, $(\forall \gamma < \beta (\varphi_\gamma)) \to \varphi_\beta$ holds. Then $\forall \beta < \alpha (\varphi_\beta)$ holds.*

Theorem 3.2.18 (Definition by transfinite induction) *Let X be a non-empty set and $\alpha \in ON$. Set $\mathcal{F}$ to be the set of all functions with domain some $\beta < \alpha$ and range contained in X. Then given any function $G : \mathcal{F} \to X$ there is a unique function $f : \alpha \to X$ such that $\forall \beta < \alpha (f(\beta) = G(f|\beta))$.*

We give an example of a definition by transfinite induction. Let $A \subset \mathbb{R}$ be a closed and bounded set. Set $A^{(0)} = A$. Let $\beta < \omega_1$. Suppose $A^{(\gamma)}, \gamma < \beta$, have been defined. If β is a limit ordinal, we define $A^{(\beta)} = \cap_{\gamma < \beta} A^{(\gamma)}$. If $\beta = \gamma + 1$ is a successor ordinal, then we define $A^{(\beta)} = (A^{(\gamma)})'$, the set of all accumulation points of $A^{(\gamma)}$. Sets A^α are called the *Cantor-Bendixson derivatives* of A. Note that for $\gamma < \beta < \omega_1$, $A^\beta \subset A^\gamma$.

Exercise 3.2.19 For every successor ordinal $\alpha = \beta + 1$, show that there is a closed and bounded subset A of $\mathbb{R}$ such that $A^{(\beta)} \neq \emptyset$ and $A^{(\beta+1)} = \emptyset$.

We close this section by giving a *canonical well-order* on the class $ON \times ON$. This will be used to show that AC is essential for the cardinal arithmetic. For $\alpha, \beta, \gamma, \delta \in ON$, we define

$$(\alpha, \beta) \prec (\gamma, \delta) \Leftrightarrow \text{ either } \max\{\alpha, \beta\} < \max\{\gamma, \delta\}$$
$$\text{or } \max\{\alpha, \beta\} = \max\{\gamma, \delta\} \text{ and } \alpha < \gamma$$
$$\text{or } \max\{\alpha, \beta\} = \max\{\gamma, \delta\}, \alpha = \gamma \text{ and } \beta < \delta.$$

We leave it for the reader to verify that this defines a well-order on the class $ON \times ON$.

Exercise 3.2.20 Show that for ordinals α, β and γ,

$$(\beta, \gamma) \prec (0, \alpha) \Leftrightarrow \beta, \gamma < \alpha.$$

3.3 Cardinal Numbers

A *cardinal number* is an ordinal number κ such that for every ordinal number $\alpha < \kappa$ there is no bijection $f : \alpha \to \kappa$. Sometimes cardinal numbers are also called *initial ordinals*. For instance, $0, 1, 2, \ldots$, ω_0 and ω_1 are cardinal numbers. Further, ω_0 is the first infinite cardinal and ω_1 the first uncountable cardinal.

Theorem 3.3.1 *An infinite cardinal number is a limit ordinal number.*

Proof Let κ be an infinite cardinal. By Theorem 3.2.9, $\kappa = \beta + n$ where β is a non-zero and limit and $n \in \omega_0$. If $0 < n$, then there is a bijection from β onto κ by Proposition 2.1.33. Since $\beta < \kappa$, we get a contradiction. ∎

Let X be any set. Assume WOP. Then there is a well order $\leq$ on X. By Theorem 3.2.7 there is an ordinal number α such that $X \sim \alpha$. Now consider

$$A = \{\beta \in ON : \beta \leq \alpha \ \& \ \text{there is a bijection} \ f : \alpha \to \beta\}.$$

Then A is a non-empty set of ordinal numbers. Its least element is obviously a cardinal number. We denote the least element of A by $|X|$ and call it the *cardinal number* or simply the *cardinality* of the set X.

Exercise 3.3.2 Check that $|X|$ is well defined.

The following facts are quite easily seen.

Proposition 3.3.3 *Let X and Y be two sets. Then*

1. $|X| = |Y|$ *if and only if there is a bijection from X to Y.*
2. $|X| < |Y|$ *if and only if there is an injection from X to Y but there is no bijection from X to Y.*

Cantor proved that for every set X, $|X| < |\mathcal{P}(X)|$. This implies that the class of all cardinal numbers is unbounded in ON, i.e., there is no largest cardinal number. Indeed, we have the following result.

Proposition 3.3.4 *Let $\alpha \in ON$ and $\{\kappa_\beta : \beta < \alpha\}$ a set of cardinal numbers. Then there is a cardinal number κ such that for all $\beta < \alpha$, $\kappa_\beta < \kappa$.*

Proof Clearly, $\gamma = \cup_{\beta < \alpha} \kappa_\beta \in ON$. By Cantor's theorem, there is a cardinal $\kappa > \gamma$. This works. ∎

For any cardinal number κ, κ^+ will denote the least cardinal number greater than κ. Such cardinals are called *successor cardinals* and other cardinals are called *limit cardinals*. Note that $\omega_0^+ = \omega_1$.

We have already seen that $\omega_0 < \omega_1$ are cardinal numbers. For each ordinal number α, we define ω_α by transfinite induction as follows: Suppose $\alpha \in ON$ and $\omega_\beta, \beta < \alpha$, have been defined such that for $\gamma < \beta < \alpha$, $\omega_\gamma < \omega_\beta$. If $\alpha = \beta + 1$ is a successor ordinal, then we define

$$\omega_{\beta+1} = \{\gamma \in ON : |\gamma| \leq |\omega_\beta|\}.$$

If α is a limit ordinal, we define $\omega_\alpha = \sup_{\beta < \alpha} \omega_\beta = \cup_{\beta < \alpha} \omega_\beta$.

Theorem 3.3.5 *For each $\alpha \in ON$, ω_α is a cardinal number. Moreover, if $\alpha = \beta + 1$, then $\omega_\alpha = \omega_\beta^+$.*

3.3 Cardinal Numbers

Proof Let $\alpha \in ON$ be such that for all $\beta < \alpha$, ω_β is a cardinal number.

Suppose $\alpha = \beta + 1$ for some $\beta \in ON$. Then $\omega_\beta \in \omega_\alpha$. Hence, $\omega_\beta \subset \omega_\alpha$. Therefore, $|\omega_\beta| \leq |\omega_\alpha|$. If possible, suppose $|\omega_\beta| \geq |\omega_\alpha|$. Then $\omega_\alpha \in \omega_\alpha$ which is a contradiction.

Next assume that α is a limit ordinal. Clearly, $|\omega_\beta| < |\omega_\alpha|$ for all $\beta < \alpha$. If $\gamma < \omega_\alpha$, then $\gamma < \omega_\beta$ for some $\beta < \alpha$. In particular, $|\gamma| < |\omega_\alpha|$. It follows that ω_α is a cardinal number. ∎

Clearly for every $\alpha \in ON$, $\alpha \leq \omega_\alpha$. We have the following important corollary now.

Corollary 3.3.6 *An infinite ordinal number κ is a cardinal number if and only if $\kappa = \omega_\alpha$ for some $\alpha \in ON$.*

Exercise 3.3.7 Give an example of an ordinal α such that $\omega_\alpha = \alpha$.

When one is interested in ω_α as a cardinal number only, one writes $\aleph_\alpha$ for ω_α. We call $\aleph_\alpha$'s *alephs*. (We have used $\mathbb{N}$ for ω_0 till we introduced the notation ω_0. This notation is used when one is doing arithmetic and elements of $\mathbb{N}$ are treated as natural numbers.)

In Chap. 2, we have done sufficient work to introduce cardinal arithmetic. Let X and Y be two sets of cardinality λ and μ respectively.

(a) Without any loss of generality we assume that $X \cap Y = \emptyset$. Define $\lambda + \mu = |X \cup Y|$.
(b) In general, assume that $\{X_i : i \in I\}$ is a family of pairwise disjoint sets and $|X_i| = \lambda_i, i \in I$. We define $\sum_{i \in I} \lambda_i = |\cup_{i \in I} X_i|$.
(c) We define $\lambda \cdot \mu = |X \times Y|$.
(d) In general, assume that $\{X_i : i \in I\}$ is a family of sets and $|X_i| = \lambda_i, i \in I$. We define $\Pi_{i \in I} \lambda_i = |\times_{i \in I} X_i|$.
(e) We define $\lambda^\mu = |X^Y|$.

These definitions do not depend on the choice of sets. (See Exercise 2.1.6.) We rewrite several observations that we made in Chap. 2 in terms of cardinal numbers now.

Theorem 3.3.8 *1. Since there is a bijection between $\{0, 1\}^\mathbb{N}$ and $\mathbb{N}^\mathbb{N}$, $2^{\aleph_0} = \aleph_0^{\aleph_0}$. From here it easily follows that for any natural number k, $2^{\aleph_0} = k^{\aleph_0} = \aleph_0^{\aleph_0}$.*
2. Since for any infinite set X, $|X| = |X \times \{0, 1\}|$, for any infinite cardinal λ, $\lambda = \lambda + \lambda$.
3. Since for any infinite set X, $|X| = |X \times X|$, for any infinite cardinal λ, $\lambda = \lambda \cdot \lambda$.
4. Let X, Y, Z be sets with $Y \cap Z = \emptyset$. Then $|X^Y \times X^Z| = |X^{Y \cup Z}|$. Hence for cardinal numbers λ, μ and ν, $\lambda^\mu \cdot \lambda^\nu = \lambda^{\mu+\nu}$.
5. Since for sets X, Y and Z, $|(X^Y)^Z| = |X^{Y \times Z}|$, for cardinals λ, μ and ν, $(\lambda^\mu)^\nu = \lambda^{\mu \cdot \nu}$.

We write $|\mathbb{R}| = \mathfrak{c}$ and call it the *continuum*. From the theory of simple continued fraction, it is well known that the map

$$(n_0, n_1, n_2, n_3, \ldots) \to n_0 + \cfrac{1}{(n_1+1)+} \cfrac{1}{(n_2+1)+} \cfrac{1}{(n_3+1)+}$$

from $\mathbb{N}^{\mathbb{N}}$ to $\mathbb{R}$ is one-to-one and onto the set of all positive irrational numbers. By (2), the set of all positive irrational numbers and the set of all irrational numbers are of the same cardinality. Using this and (1) above it follows that

$$2^{\aleph_0} = \aleph_0^{\aleph_0} = |\mathbb{R} \setminus \mathbb{Q}| = |\mathbb{R}| = \mathfrak{c}.$$

From D. König's Theorem 2.1.40 we see that if $\{\lambda_i : i \in I\}$ and $\{\mu_i : i \in I\}$ are sets of cardinals such that for each $i \in I$, $\lambda_i < \mu_i$, then $\sum_{i \in I} \lambda_i < \prod_{i \in I} \mu_i$.

Exercise 3.3.9 Let λ, μ and ν be cardinal numbers. Show the following.

1. $\lambda \cdot (\mu + \nu) = \lambda \cdot \mu + \lambda \cdot \nu$.
2. $(\lambda \cdot \mu)^{\nu} = \lambda^{\nu} \cdot \mu^{\nu}$.
3. $\lambda \leq \mu \Rightarrow \lambda + \nu \leq \mu + \nu$.
4. $\lambda \leq \mu \Rightarrow \lambda \cdot \nu \leq \mu \cdot \nu$.
5. $\lambda \leq \mu \Rightarrow \lambda^{\nu} \leq \mu^{\nu}$.
6. If not both λ and ν are zero, then $\lambda \leq \mu \Rightarrow \nu^{\lambda} \leq \nu^{\mu}$.

What is $\mathfrak{c}$, i.e., for which $\alpha \in ON$, $\mathfrak{c} = \aleph_\alpha$? This had been one of the foremost questions in twentieth century mathematics. Cantor believed that $\mathfrak{c} = \aleph_1$, i.e., $2^{\aleph_0} = \aleph_1$. This is known as the *continuum hypothesis*.

Continuum Hypothesis, (CH). $2^{\aleph_0} = \aleph_1$.

Since $\mathfrak{c} = 2^{\aleph_0}$ and $\aleph_1 = \aleph_0^+$, this question is equivalent to the following: *Let $A \subset \mathbb{R}$ be uncountable. Must then there exist a bijection from A onto $\mathbb{R}$?*

One may also ask: Is it true that for every $\alpha \in ON$, $2^{\aleph_\alpha} = \aleph_{\alpha+1}$, i.e., is $2^{\aleph_\alpha} = \aleph_\alpha^+ = \aleph_{\alpha+1}$? This is known as *Generalized continuum hypothesis*.

Generalized Continuum Hypothesis, (GCH). For every ordinal α, $2^{\aleph_\alpha} = \aleph_{\alpha+1}$ or for every infinite cardinal κ, $2^\kappa = \kappa^+$.

As one of the most remarkable results in set theory, using axioms of ZF, Gödel defined a class model L, called the class of all *constructible sets*, of ZF in which WOP and GCH hold. This implies that using the axioms of ZF one cannot disprove either WOP or CH or GCH.

As one of the most magical results ever proved in set theory, Cohen discovered a technique, called *forcing*, of building models of ZF, and constructed models of $ZF + \neg AC$ as well as models of ZFC in which $2^{\aleph_0} > \aleph_1$. Hence, using the axioms

3.3 Cardinal Numbers

of ZF one cannot prove either AC or CH either. Today the forcing techniques of Cohen and several refinements of it are the most powerful techniques to build models of ZF and prove deep results in set theory.

By transfinite induction we define $\beth_\alpha$, $\alpha \in ON$, such that

$$\beth_0 = \aleph_0,$$

$$\beth_{\beta+1} = 2^{\beth_\beta}$$

and if δ is a limit ordinal, then

$$\beth_\delta = \cup_{\beta \in \delta} \beth_\beta.$$

These are called *Beth cardinals*. By inducting on $\alpha \in ON$, it can be easily seen that

$$GCH \Leftrightarrow \forall \alpha \in ON(\beth_\alpha = \aleph_\alpha).$$

These topics are well beyond the scope of this note. Readers are strongly advised to see [1, 2] for these results.

Exercise 3.3.10 1. Show that $|\mathbb{R}| = |\mathbb{R}^k| = |\mathbb{R}^\mathbb{N}|$, where $k > 1$.
2. Show that $|\{z \in \mathbb{C} : |z| < 1\}| = |\{z \in \mathbb{C} : |z| \leq 1\}| = |\mathbb{C}|$.
3. Let $\lambda \leq \mu$ be cardinals with μ infinite. Show that $\lambda + \mu = \mu = \lambda \cdot \mu$.
4. Let μ be an infinite cardinal, $\{A_i : i \in I\}$ a family of sets each of cardinality $\leq \mu$ and $|I| = \kappa$, where $\kappa \leq \mu$. Then show that $|\cup_{i \in I} A_i| \leq \mu$.

Lemma 3.3.11 $|\mathbb{R}^\mathbb{R}| = |\{0, 1\}^\mathbb{R}|$.

Proof

$$\mathfrak{c}^\mathfrak{c} = (2^{\aleph_0})^\mathfrak{c} = 2^{\aleph_0 \cdot \mathfrak{c}} = 2^\mathfrak{c}.$$

∎

Exercise 3.3.12 Show that the set of all continuous real-valued functions on $\mathbb{R}$ is of cardinality $\mathfrak{c}$.

Exercise 3.3.13 Show that the set of all non-empty open sets is of cardinality $\mathfrak{c}$.

Exercise 3.3.14 1. Let $\{A_\alpha : \alpha < \omega_1\}$ be a transfinite decreasing sequence of subsets of $\mathbb{R}$ such that for every limit ordinal $\alpha < \omega_1$, $A_\alpha = \cap_{\beta < \alpha} A_\beta$. Further assume that either all A_α are closed or all are open. Show that there is a $\alpha < \omega_1$ such that for all $\beta > \alpha$, $A_\beta = A_\alpha$.
2. Let $\{A_\alpha : \alpha < \omega_1\}$ be a transfinite increasing sequence of subsets of $\mathbb{R}$ such that for every limit ordinal $\alpha < \omega_1$, $A_\alpha = \cup_{\beta < \alpha} A_\beta$. Further assume that either all A_α are closed or all are open. Show that there is a $\alpha < \omega_1$ such that for all $\beta > \alpha$, $A_\beta = A_\alpha$.

(**Hint.** Let $\{A_\alpha : \alpha < \omega_1\}$ be a transfinite decreasing sequence of closed subsets of $\mathbb{R}$ such that for every limit ordinal $\alpha < \omega_1$, $A_\alpha = \cap_{\beta < \alpha} A_\beta$. Fix a countable base $\{U_n\}$ for the topology of $\mathbb{R}$. For instance, we may take $\{U_n\}$ to be an enumeration of all open intervals with rational end points.

Consider the set
$$A = \{\alpha < \omega_1 : A_\alpha \setminus A_{\alpha+1} \neq \emptyset\}.$$

Let $\alpha \in A$ and $x \in A_\alpha \setminus A_{\alpha+1}$. Since $A_{\alpha+1}$ is closed, there exists a natural number n_α such that
$$x \in U_{n_\alpha} \subset \mathbb{R} \setminus A_{\alpha+1}.$$

Then $U_{n_\alpha} \cap A_\alpha \neq \emptyset$ and $U_{n_\alpha} \cap A_{\alpha+1} = \emptyset$.

It follows that $\alpha \to n_\alpha$ is a one-to-one map from A into $\mathbb{N}$. This proves that $A \subset \omega_1$ is countable. Get a $\gamma < \omega_1$ bigger than all $\alpha \in A$. Then $A_\beta = A_\gamma$ for all $\gamma < \beta < \omega_1$.)

3.4 An Alternative Approach to Cardinal Numbers

In the last section, we defined infinite cardinal numbers as alephs, i.e., as initial ordinals. More precisely, $\{\omega_\alpha : \alpha \in ON\}$ is precisely the class of all infinite cardinals. We have the convention that when viewed as cardinal numbers we write $\aleph_\alpha$ for ω_α, $\alpha \in ON$. In this section we show the very easy interesting fact that most of the results on cardinal arithmetic of alephs don't require AC, i.e., they are theorems in ZF. This is where we make the crucial use of the canonical ordering on ON.

The following is a result in ZF.

Proposition 3.4.1 *For every ordinal α, $|\omega_\alpha \times \omega_\alpha| = |\omega_\alpha|$, i.e., $\aleph_\alpha \cdot \aleph_\alpha = \aleph_\alpha$.*

Proof We prove it by induction on α. The result is true for $\alpha = 0$ because $|\omega_0 \times \omega_0| = |\omega_0|$ is true in ZF.

Assume that $\alpha > 0$ and the result is true for all $\beta < \alpha$. We equip $\omega_\alpha \times \omega_\alpha$ with the canonical well order $\prec$.

Take ordinals $\beta, \gamma < \omega_\alpha$. Then (β, γ) has no more than $|\beta + 1| \cdot |\gamma + 1| < \aleph_\alpha$ many predecessors in $(\omega_\alpha \times \omega_\alpha, \prec)$ by induction hypothesis. Hence, the order type of $(\omega_\alpha \times \omega_\alpha, \prec)$ is $\leq \omega_\alpha$. This implies that $\aleph_\alpha \cdot \aleph_\alpha \leq \aleph_\alpha$. Clearly, $\aleph_\alpha \cdot \aleph_\alpha \geq \aleph_\alpha$, The result follows. ∎

Since $\aleph_\alpha \leq \aleph_\alpha + \aleph_\alpha$, we have the following result without using AC.

Corollary 3.4.2 *For ordinals α, β,*
$$\aleph_\alpha + \aleph_\alpha = \aleph_\alpha \cdot \aleph_\alpha = \aleph_\alpha,$$

and

$$\aleph_\alpha + \aleph_\beta = \aleph_\alpha \cdot \aleph_\beta = \max\{\aleph_\alpha, \aleph_\beta\}.$$

Remark 3.4.3 We have not used WOP so far in this section. Hence, the usual cardinal arithmetic for alephs holds without using AC. In Theorem 3.2.7, we showed that every well-ordered set is order isomorphic to an ordinal number. Hence, under AC, for every infinite set X, there is an aleph of the same cardinality as that of X. Since the cardinal arithmetic with alephs does not require AC, we could have avoided the Sect. 2.4 completely. However, results proved in the Sect. 2.4 used Zorn's lemma which is more accessible, particularly to non-set theorists, because it avoids ordinals completely. Hence, we have developed cardinal arithmetic using ZL also.

3.5 Cardinal Arithmetic and Axiom of Choice

In this section, we show that AC is needed to develop the cardinal arithmetic. *Unless otherwise stated, results in this section do not use AC.* Also, for sets X, Y, $|X| \leq |Y|$ ($|X| = |Y|$) will mean that there is an injection (respectively bijection) $f : X \to Y$.

Theorem 3.5.1 (Hartog's Theorem) *For every set A there is a least aleph $\aleph_A$ such that $|\aleph_A| \not\leq |A|$, i.e., there is no one-to-one map from $\aleph_A$ to A.*

Proof Take any set A. Consider

$$\aleph_A = \{\beta \in ON : |\beta| \leq |A|\}.$$

Main Observation. $\aleph_A$ is a set.

This is true because for any $\beta \in ON$,

$$\beta \in \aleph_A \Leftrightarrow \exists f \in \mathcal{P}(\beta \times A)(f \text{ is an injection from } \beta \text{ to } A).$$

By replacement axiom $\aleph_A$ is a set.

Clearly, if $\beta \in \aleph_A$ then every ordinal $\gamma < \beta$ is in $\aleph_A$. Hence, $\aleph_A$ is an ordinal. If $\beta \in \aleph_A$, then $|\beta| \leq |A|$. This implies that $\aleph_A$ is the least aleph such that $|\aleph_A| \not\leq |A|$. If $|\aleph_A| \leq |A|$, the $\aleph_A \in \aleph_A$, contradicting the regularity axiom. ∎

Theorem 3.5.2 *If for any two sets A and B, either $|A| \leq |B|$ or $|B| \leq |A|$, then WOP holds.*

Proof Let A be any infinite set. Since $|\aleph_A| \not\leq |A|$, by our hypothesis, $|A| \leq \aleph_A$. This implies that A can be well-ordered. ∎

Theorem 3.5.3 *For any infinite set A and any ordinal α,*

$$|A| + \aleph_\alpha = |A| \cdot \aleph_\alpha \Rightarrow |A| \leq \aleph_\alpha \vee \aleph_\alpha \leq |A|.$$

Proof By our hypothesis, there exist two disjoint sets A_1 and B_1 such that $|A_1| = |A|$, $|B_1| = \aleph_\alpha$ and $A \times \omega_\alpha = A_1 \cup B_1$.

Suppose there exists $a \in A$ such that $(a, \beta) \in A_1$ for every $\beta \in \omega_\alpha$. Then $\aleph_\alpha \leq |A_1| = |A|$. Or for every $a \in A$, there exists least $\beta_a < \omega_\alpha$ such that $(a, \beta_a) \in B_1$. This implies that $|A| \leq \aleph_\alpha$. ∎

Since $\aleph_A \nleq |A|$, we have

Corollary 3.5.4 *Let A be an infinite set. Then $|A| + \aleph_A = |A| \cdot \aleph_A$ implies that $|A|$ is an aleph.*

Theorem 3.5.5 *If for every infinite set A, $|A \times A| = |A|$, then WOP holds.*

Proof Let A be an infinite set. Sufficient to prove that $|A| \leq \aleph_A$. By Corollary 3.5.4, it is sufficient to prove that

$$|A| + \aleph_A = |A| \cdot \aleph_A,$$

Clearly, $|A| + \aleph_A \leq |A| \cdot \aleph_A$. Hence, we need to prove $|A| + \aleph_A \geq |A| \cdot \aleph_A$. By our hypothesis,

$$|A| + \aleph_A = (|A| + \aleph_A) \cdot (|A| + \aleph_A).$$

It is not hard to prove that

$$(|A| + \aleph_A) \cdot (|A| + \aleph_A) \geq |A|.\aleph_A.$$

∎

3.6 More on Cardinals and Ordinals

Let $\alpha \in ON$. We define

$$cf(\alpha) = \min\{\beta \leq \alpha : \exists f : \beta \to \alpha (\sup f(\beta) = \alpha)\}.$$

$cf(\alpha)$ is called the *cofinality* of α. It equals the least ordinal $\beta \leq \alpha$ such that there is a function $f : \beta \to \alpha$ satisfying $\sup\{f(\delta) : \delta < \beta\} = \alpha$. Such a function f is called *unbounded*.

Example 3.6.1 $cf(\omega_0) = \omega_0$, $cf(\omega_1) = \omega_1$ and $cf(\omega_{\omega_0}) = \omega_0 < \omega_{\omega_0}$.

Let $\gamma < \beta < \alpha$ be ordinals and $f : \beta \to \alpha$, $g : \gamma \to \beta$ unbounded. Then $f \circ g : \gamma \to \alpha$ is unbounded. This observation together with the trivial fact that a bijection $f : \gamma \to \beta$ is unbounded, immediately gives us the following result.

3.6 More on Cardinals and Ordinals

Proposition 3.6.2 *For every ordinal α,*

(a) $cf(\alpha)$ is a cardinal number.
(b) $cf(cf(\alpha)) = cf(\alpha)$.

An infinite cardinal number κ is called *regular* if $cf(\kappa) = \kappa$. Otherwise, κ is called a *singular cardinal*. We see that ω_0 and ω_1 are regular cardinals and ω_{ω_0} is a singular cardinal. Also, if α is a limit ordinal, then $cf(\omega_\alpha) = cf(\alpha)$.

Note the trivial fact that if there is an unbounded function $g : \gamma \to \alpha$, then there is an unbounded function $f : \beta \to \alpha$ whenever $\gamma < \beta < \alpha$.

Example 3.6.3 Show that for any ordinal α, $cf(\alpha)$ is a regular cardinal.

Theorem 3.6.4 *For every infinite cardinal κ, κ^+ is regular.*

Proof If possible, suppose κ^+ is singular. Then there is an unbounded function $f : \kappa \to \kappa^+$. Since f is unbounded, $\cup_\alpha f(\alpha) = \kappa^+$. But $|\cup_\alpha f(\alpha)| = \kappa$. This contradiction proves our result. ∎

Lemma 3.6.5 *If κ is an infinite cardinal and $cf(\kappa) \leq \lambda$, then $\kappa < \kappa^\lambda$.*

Proof Let $g : \kappa \to \kappa^\lambda$ be any function. It is sufficient to prove that g is not a surjection. Since $cf(\kappa) \leq \lambda$, there is an unbounded function $f : \lambda \to \kappa$. Take any $\alpha < \lambda$. The set
$$B_\alpha = \{g(\mu)(\alpha) \in \kappa : \mu < f(\alpha)\}$$
is of cardinality less than κ. Hence $A_\alpha = \kappa \setminus B_\alpha \neq \emptyset$. Consider the function $h : \lambda \to \kappa$ such that $h(\alpha) = \min(A_\alpha)$ for each $\alpha < \lambda$. Then h does not belong to the range of g and our proof is complete. ∎

Proposition 3.6.6 *If λ is an infinite cardinal, then $cf(2^\lambda) > \lambda$.*

Proof Since λ is infinite,
$$(2^\lambda)^\lambda = 2^{\lambda \cdot \lambda} = 2^\lambda.$$

By taking $\kappa = 2^\lambda$ in the last lemma, we get our result. ∎

Recall that in Chap. 2 in the section on linearly ordered sets, we had introduced the notion of closed and open subsets of a linearly ordered set.

Proposition 3.6.7 *A subset C of ω_1 is closed if and only if for every limit ordinal $\alpha < \omega_1$, whenever $C \cap \alpha$ is unbounded, $\alpha \in C$.*

Proof Note that a non-empty interval $I \subset \omega_1$ with left end point α is open if and only if whenever $\alpha < \omega_1$ is a limit ordinal and the left end point of I, $\alpha \notin I$. The sentence on the right of 'if and only if' is equivalent to saying that $\omega_1 \setminus C$ is a union of open intervals. This completes the proof. ∎

A closed and unbounded subset of ω_1 is called a *CUB*.

Proposition 3.6.8 *If $A, B \subset \omega_1$ are CUBs, then $A \cap B$ is also a CUB.*

Proof Since the union of a family of open subsets of a linearly ordered set is clearly open, by De Morgan's theorem, the intersection of a family of closed subsets of a linearly ordered set is closed. Hence, we only need to show that $A \cap B$ is unbounded.

Take any $\beta < \omega_1$. Since A is unbounded, let α_0 be the least countable ordinal greater than β that belongs to A. Since B is unbounded, let $\beta_0 > \alpha_0$ be the least countable ordinal in B. Then let $\alpha_1 > \beta_0$ be the least countable ordinal in A. Next let $\beta_1 > \alpha_1$ be the least countable ordinal in B. Proceeding similarly, we get

$$\beta < \alpha_0 < \beta_0 < \alpha_1 < \beta_1 < \cdots$$

such that for each n, $\alpha_n \in A$ and $\beta_n \in B$. Clearly, $\sup \alpha_n = \sup_n \beta_n = \gamma$, say. Since A and B are closed, $\gamma \in A \cap B$. ∎

Proposition 3.6.9 *If $\{A_n\}$ is a sequence of CUBs, then $\cap_n A_n$ is a CUB.*

Proof Since $|\mathbb{N} \times \mathbb{N}| = \mathbb{N}$, there is a partition $\mathbb{N} = \cup_n I_n$ of $\mathbb{N}$ into infinite sets I_n. We need only to show that $\cap_n A_n$ is unbounded.

Take any $\beta < \omega_1$. Arguing as in the proof of the last proposition, there is a sequence

$$\beta < \alpha_0 < \alpha_1 < \cdots < \alpha_k < \cdots$$

of countable ordinals such that whenever $k \in I_n$, $\alpha_k \in A_n$. Clearly, for every n, $\sup_{k \in I_n} \alpha_k$ are all equal and equal to, say γ. Since each A_n is closed, $\gamma \in \cap_n A_n$. ∎

We close this chapter with two technical lemmas (that follow easily from cardinal arithmetic and transfinite induction) that are useful in several branches of mathematics including set theory. Let A be a non-empty set and κ a cardinal. A function $A^\kappa \to A$ will be called a *κ-ary operation* on A. If $\kappa < \omega_0$, then f will be called a *finite ary operation* on A and if $\kappa \leq \omega_0$ it will be called a *countable ary operation* on A. Let $\mathcal{A}$ be a class of operations of arbitrary arities on a non-empty set A and $B \subset A$. We shall say that B is closed under $\mathcal{A}$ if for every $f \in \mathcal{A}$, say of κ-arity, $f(B^\kappa) \subset B$.

Lemma 3.6.10 *Let κ be an infinite cardinal, $A \neq \emptyset$, $B \subset A$ of cardinality $\leq \kappa$ and $\mathcal{A}$ a non-empty set of finite ary operations on A of cardinality $\leq \kappa$. Then the smallest subset C of A containing B and closed under $\mathcal{A}$ is of cardinality $\leq \kappa$.*

Proof By induction on natural numbers, we define a sequence $\{C_n\}$ of subsets of A as follows: $C_0 = B$ and

$$C_{n+1} = C_n \cup \cup_k \cup \{f(C_n^k) : f \in \mathcal{A} \ \& \ f \text{ is } k - \text{ary}\}.$$

By induction on n and using routine cardinal arithmetic, it is easily seen that for each n, $|C_n| \leq \kappa$. Therefore, $|C| \leq \kappa$. Further C is the smallest subset of A containing B which is closed under $\mathcal{A}$. ∎

Lemma 3.6.11 *Let κ be an uncountable cardinal, $A \neq \emptyset$, $B \subset A$ of cardinality $\leq \kappa$ and $\mathcal{A}$ a non-empty set of countable ary operations on A of cardinality $\leq \kappa$. Then the smallest subset C of A containing B and closed under $\mathcal{A}$ is of cardinality $\leq \kappa$.*

Proof By transfinite induction, we define a transfinite sequence $\{C_\alpha : \alpha < \omega_1\}$ of subsets of A as follows:

$$C_0 = B,$$

$$C_\alpha = \cup_{\beta < \alpha} C_\beta,$$

if $\alpha < \omega_1$ is a limit ordinal and

$$C_{\alpha+1} = C_\alpha \cup \cup_{k \leq \omega_0} \cup \{f(C_\alpha^k) : f \in \mathcal{A} \ \& \ f \ \text{is} \ k-\text{ary}\}.$$

Set $C = \cup_{\alpha < \omega_1} C_\alpha$.

By transfinite induction on $\alpha < \omega_1$ and using routine cardinal arithmetic, it is easily seen that for each α, $|C_\alpha| \leq \kappa$. Since $\kappa \geq \omega_1$, $|C| \leq \kappa$. Further C is the smallest subset of A containing B which is closed under $\mathcal{A}$. ∎

References

1. T. Jech, *Set Theory*, Springer Monographs in Mathematics, 3rd edn. (Springer, New York, 2002)
2. K. Kunen, *Set Theory: An Introduction to Independence Proofs* (North-Holland Publishing Company, 1980)

Chapter 4
Applications in Other Branches of Mathematics

In this chapter, we shall give applications of set theory to other branches of mathematics. We shall assume that the reader is familiar with the subjects and give only preliminary definitions needed to understand the statements and proofs of the applications that we shall be presenting.

4.1 Analysis

Let I be an infinite set and for each $i \in I$ let a_i be a non-negative real number. We define
$$\sum_i a_i = \sup\{\sum_{i \in F} a_i : F \subset I \text{ finite}\}.$$

Note that this extends the usual definition of the sum $\sum_{j \in J} a_j$ when each $a_j \geq 0$ and J countable to uncountable sum of non-negative real numbers.

Proposition 4.1.1 *Let I be uncountable and $\{a_i : i \in I\}$ a family of positive real numbers. Then $\sum_i a_i = +\infty$. Equivalently, if each $a_i \geq 0$ and $\sum_i a_i < +\infty$, then $\{i \in I : a_i > 0\}$ is countable.*

Proof For each $n \geq 1$, let $I_n = \{i \in I : a_i > 1/n\}$. Since each $a_i > 0$, $I = \cup_n I_n$. Since I is uncountable there is a $n \geq 1$ such that I_n is infinite. Then $\sup\{\sum_{i \in F} a_i : F \subset I_n \text{ finite}\} = +\infty$. Our result follows. ∎

Remark 4.1.2 Let I be an infinite set. Then
$$D = \{F \subset I : F \text{ finite}\}$$

is a directed set with respect to the inclusion $\subset$. Take a Banach space X and $a_i \in X$, $i \in I$. Then

$$\{\sum_{i \in F} a_i \in X : F \subset I \text{ finite}\}$$

is a net in X. Now we can define

$$\sum_i a_i = \lim_{F \in D} \sum_{i \in F} a_i.$$

The limit exists would imply that all but countably many terms of the net would be zero.

A complex number is called *algebraic* if it is a root of a non-zero polynomial in a single variable with integer coefficients. For instance, every rational number, $\sqrt{2}, i$, and $3 + 2i$ are algebraic numbers. It is easy to see that a complex number is algebraic if and only if it is a root of a non-zero polynomial in a single variable with rational coefficients. A complex number which is not algebraic is called *transcendental*.

Proposition 4.1.3 *The set of all algebraic numbers is countable.*

Proof We know that the set $\mathbb{Z}^{<\mathbb{N}}$ of all finite sequences of integers is countable. This implies that the set of all non-zero polynomials in a single variable with integer coefficients is countable. Let $\{f_n : n \in \mathbb{N}\}$ be an enumeration of all non-zero polynomials in a single variable with integer coefficients. For each $n \in \mathbb{N}$, let Z_n denote the set of all roots of f_n. Then each Z_n is finite and $\cup_n Z_n$ is the set of all algebraic numbers. Our result follows from the fact that the union of a sequence of countable sets is countable. ∎

Since the set of all real numbers is uncountable, we have the following result.

Corollary 4.1.4 *There is a real transcendental number.*

Remark 4.1.5 This was the first proof (due to Cantor) of the existence of a transcendental number. Till then no real or complex number was shown to be transcendental. Since Cantor only showed the existence of a transcendental number without giving any example, the proof was not accepted by some prominent mathematicians of that time.

Remark 4.1.6 Since we can define linear orders on $\mathbb{Z}^{<\mathbb{N}}$ and $\mathbb{C}$ (e.g., the lexicographic orders on $\mathbb{Z}^{<\mathbb{N}}$ and on $\mathbb{R} \times \mathbb{R}$, respectively), the above results can be proved without using AC.

Exercise 4.1.7 Show that the set of all real transcendental numbers is of cardinality $\mathfrak{c}$.

Proposition 4.1.8 *Every non-empty closed set in $\mathbb{R}$ with no isolated points is of cardinality $\mathfrak{c}$.*

4.1 Analysis

Proof Let C be a non-empty closed set in $\mathbb{R}$ with no isolated points. By induction on the length $|s|$ of $s \in \{0, 1\}^{<\mathbb{N}}$, we define a non-empty open set U_s in $\mathbb{R}$ satisfying the following conditions:

1. $U_e = \mathbb{R}$ where e is the empty sequence of 0s and 1s.
2. $U_s \cap C \neq \emptyset$ for every $s \in \{0, 1\}^{<\mathbb{N}}$.
3. For every $s \in \{0, 1\}^{<\mathbb{N}}$ of length at least 1,

$$\text{diameter}(U_s) < 2^{-|s|}.$$

4. For every $s \in \{0, 1\}^{<\mathbb{N}}$ and $\epsilon = 0$ or 1,

$$\overline{U_{s\frown\epsilon}} \subset U_s.$$

5. $(s \neq t \wedge |s| = |t|) \Rightarrow \overline{U_s} \cap \overline{U_t} = \emptyset.$

Assuming that a system of open sets $\{U_s : s \in \{0, 1\}^{<\mathbb{N}}\}$ satisfying above conditions have been defined, we complete the proof first. Take any $\alpha \in \{0, 1\}^{\mathbb{N}}$. For each $k \in \mathbb{N}$, consider

$$F_k = \overline{U_{\alpha|k}} \cap C.$$

Then $\{F_k\}$ is a decreasing sequence of non-empty closed sets in $\mathbb{R}$ of diameters converging to 0 as $k \to \infty$. By the Cantor intersection theorem [1, Proposition 2.1.29], $\cap_k F_k$ contains a unique point, say $f(\alpha)$.

Let $\alpha \neq \beta \in \{0, 1\}^{\mathbb{N}}$. Then there exists a k such that $\alpha|k = s \neq t = \beta|k$. By (5) above. $f(\alpha) \neq f(\beta)$. Hence, $f : \{0, 1\}^{\mathbb{N}} \to C$ is one to one. Our result follows from here.

We now define U_s, $s \in \{0, 1\}$ satisfying above conditions. Suppose $|s| = k \geq 0$ and an open set U_s with $U_s \cap C \neq \emptyset$ has been defined. Since C has no isolated points, there exist $x_0 \neq x_1 \in U_s \cap C$. Since U_s is open, for each $\epsilon = 0$ or 1, there exists an open set $U_{s\frown\epsilon}$ of diameter less than 2^{-k-1} such that

$$x_\epsilon \in U_{s\frown\epsilon} \subset \overline{U_{s\frown\epsilon}} \subset U_s$$

and

$$\overline{U_{s\frown 0}} \cap \overline{U_{s\frown 1}} = \emptyset.$$

It is fairly routine to check that the system of open sets $\{U_s : s \in \{0, 1\}^{<\mathbb{N}}\}$ so defined satisfy above conditions. ∎

Exercise 4.1.9 Show that the function $f : \{0, 1\}^{\mathbb{N}} \to C$ defined in the above proof is continuous when $\{0, 1\}^{\mathbb{N}}$ is equipped with the product of discrete topologies on $\{0, 1\}$. In particular, $f : \{0, 1\}^{\mathbb{N}} \to C$ is an embedding.

Remark 4.1.10 The proof of Proposition 4.1.8 goes through verbatim and shows that every non-empty perfect set in a complete metric space is of cardinality at least

c. The proof also shows the usefulness of the Cantor intersection theorem. Later in this section we shall use same technique to prove that every uncountable Borel set in a complete separable metric space is of cardinality c. Indeed this technique is very useful and has been used to prove many deep results.

We now give some very beautiful and important applications of the characterization of the order type of $\mathbb{Q}$ given in Theorem 2.5.13. Let $f : \mathbb{R} \to \mathbb{R}$ be any function. If $\lim_{x \to \infty} f(x) = t \in \mathbb{R}$, we shall write $f(\infty) = t$. Similarly, if $\lim_{x \to -\infty} f(x) = s \in \mathbb{R}$, we write $f(-\infty) = s$.

Theorem 4.1.11 (Cantor) *Let $C \subset \mathbb{R}$ be a non-empty perfect set not containing any non-empty open interval. Then there is a continuous, non-decreasing function $f : \mathbb{R} \to [0, 1]$ such that $f(-\infty) = 0$, $f(\infty) = 1$, $f|C$ is strictly increasing and f is constant on each open interval disjoint from C.*

Proof Let $\mathcal{I}$ be the family of pairwise disjoint, non-empty open intervals of $\mathbb{R}$ such that $\mathbb{R} \setminus C = \cup \mathcal{I}$. Since $\mathbb{R}$ is separable, $\mathcal{I}$ is countable.

For $I \neq J \in \mathcal{I}$, define $I \prec J$ if the interval I is to the left of J. This is clearly a strict linear order on $\mathcal{I}$. Let $I = (a, b) \prec (c, d) = J \in \mathcal{I}$. If there is no open interval $I' \in \mathcal{I}$ between I and J, then $(b, c) \subset C$ contradicting our hypothesis.

Suppose C is bounded below and $-\infty < a = \inf C$. Then $(-\infty, a)$ is the first element of $\mathcal{I}$. Similarly, if C is bounded above and $\sup C = b < \infty$, then (b, ∞) is the last element of $\mathcal{I}$. Let $\mathcal{J}$ be obtained from $\mathcal{I}$ by removing the first and last elements of $\mathcal{I}$ (if they exist) from it. Then $(\mathcal{J}, \prec)$ is a non-empty, countable, dense linearly ordered set with no first and no last element.

Let $D \subset (0, 1)$ be a countable dense subset of $(0, 1)$. By Theorem 2.5.13, there is an order isomorphism $H : (\mathcal{J}, \prec) \to D$. Define $h : \mathbb{R} \setminus C \to [0, 1]$ by

$$h(t) = \begin{cases} 0 & \text{if } t < \inf C \\ 1 & \text{if } t > \sup C \\ H(I) & \text{if } t \in I \in \mathcal{J}. \end{cases}$$

For $x \in \mathbb{R}$, set

$$f(x) = \sup\{h(t) : t \leq x \ \& \ t \in \mathbb{R} \setminus C\}.$$

It is not hard to check that f has all the desired properties. ∎

The above theorem was essentially proved by Cantor except that he took $C \subset [0, 1]$ and containing 0 and 1. Remarkably this together with the following remark of Cantor gives a large class of probability measures on $\mathbb{R}$ singular with respect to the Lebesgue measure.

Remark 4.1.12 By varying D over all countable dense subsets of $(0, 1)$ and $H : (\mathcal{J}, \prec) \to D$ over all order isomorphisms, we get all continuous, non-decreasing surjections $f : \mathbb{R} \to [0, 1]$ such that $f|C$ is strictly increasing, f is constant on each open interval disjoint from C, $f(-\infty) = 0$ and $f(\infty) = 1$. This gives us a technique to build all non-decreasing, continuous functions $f : \mathbb{R} \to [0, 1]$ such

4.1 Analysis 71

that $f(-\infty) = 0$, $f(\infty) = 1$ and there is a perfect set C such that $f|C$ is strictly increasing and f is constant on each non-empty open interval disjoint from C. The importance of this will be seen in Section 5 on measure theory in this chapter.

Theorem 4.1.13 *Let C_1 and C_2 be two bounded perfect subsets of $\mathbb{R}$ none containing a non-empty open interval. Then there is an order isomorphism $f : \mathbb{R} \to \mathbb{R}$ such that $f(C_1) = C_2$.*

Proof Let $\mathcal{I}_1$ and $\mathcal{I}_2$ be the families of pairwise disjoint, non-empty open intervals of $\mathbb{R}$ such that $\mathbb{R} \setminus C_i = \cup \mathcal{I}_i$, $i = 1, 2$. Since $\mathbb{R}$ is separable, both $\mathcal{I}_1$ and $\mathcal{I}_2$ are countable.

Let $a_i = \inf C_i$ and $b_i = \sup C_i$, $i = 1, 2$. Since C_i is non-empty and compact, a_i, b_i are reals and both belong to C_i, $i = 1, 2$.

Set $\mathcal{J}_i = \mathcal{I}_i \setminus \{(-\infty, a_i), (b_i, \infty)\}$, $i = 1, 2$. Note that every interval in $\mathcal{J}_1$ and $\mathcal{J}_2$ are bounded.

Fix $i = 1, 2$. For intervals $I \neq J \in \mathcal{J}_i$, define $I \prec_i J$ if the interval I is to the left of J. This is clearly a strict linear order on $\mathcal{J}_i$. From the argument contained in the proof of Theorem 4.1.11, there is an order isomorphism $H : (\mathcal{J}_1, \prec_1) \to (\mathcal{J}_2, \prec_2)$.

For each $I \in \mathcal{J}_1$, fix an increasing function f_I from I onto $H(I)$. Also, fix increasing functions $f_{-\infty}$ from $(-\infty, a_1)$ onto $(-\infty, a_2)$ and f_∞ from (b_1, ∞) onto (b_2, ∞).

Set $f : \mathbb{R} \setminus C_1 \to \mathbb{R} \setminus C_2$ equal to

$$f = f_{-\infty} \cup \cup_{I \in \mathcal{J}_1} f_I \cup f_\infty.$$

Then $f : \mathbb{R} \setminus C_1 \to \mathbb{R} \setminus C_2$ is an order isomorphism.

Since C_1 and C_2 contain no non-empty open interval $\mathbb{R} \setminus C_1$ and $\mathbb{R} \setminus C_2$ are dense in $\mathbb{R}$. For $x \in C_1$, we define

$$f(x) = \sup\{f(y) : y < x \ \& \ y \in \mathbb{R} \setminus C_1\}.$$

Since $\mathbb{R}$ satisfies the lub axiom, above supremum exists. It is not hard to prove that $f : \mathbb{R} \to \mathbb{R}$ is an order isomorphism such that $f(C_1) = C_2$. ∎

Remark 4.1.14 Since an order isomorphism from $\mathbb{R}$ onto itself is a homeomorphism, f obtained in the last theorem is a homeomorphism.

Now we give a very beautiful application of set theory in complex analysis.

Proposition 4.1.15 *Let $\mathcal{F}$ be a family of entire functions such that for every complex number $z \in \mathbb{C}$, $\{f(z) : f \in \mathcal{F}\}$ is finite. Then $\mathcal{F}$ is finite.*

Proof Let $\{f_n\}$ be an infinite sequence of distinct entire functions. For natural numbers $n < m$, define

$$A_{n,m} = \{z \in \mathbb{C} : f_n(z) = f_m(z)\}.$$

Since $f_n \neq f_m$, $A_{n,m}$ is countable. Hence, $A = \cup_{n<m} A_{n,m}$ is countable. Take any $z_0 \in \mathbb{C} \setminus A$. Then whenever $n \neq m$, $f_n(z_0) \neq f_m(z_0)$. In particular, $\{f_n(z_0) : n \in \mathbb{N}\}$ is infinite. Our result is proved. ∎

Next let $\mathcal{F}$ be a family of entire functions such that for every $z \in \mathbb{C}$, $\{f(z) : f \in \mathcal{F}\}$ is countable. Must then $\mathcal{F}$ be countable? This question was asked by Wetzel in Ann Arbor Problem Book. Erdös [2] stunned people by showing that under continuum hypothesis (CH) the answer is no and under the negation of continuum hypothesis ($\neg CH$) the answer is yes.

Theorem 4.1.16 (Erdös) *The following two statements are equivalent in ZFC:*

1. $\neg CH$.
2. *Whenever $\mathcal{F}$ is a family of entire functions such that for every $z \in \mathbb{C}$, $\{f(z) : f \in \mathcal{F}\}$ is countable, $\mathcal{F}$ is countable.*

Proof Assume that $\aleph_1 < \mathfrak{c}$. Let $\{f_\alpha : \alpha < \omega_1\}$ be a family of distinct entire functions. For $\alpha < \beta < \omega_1$, set

$$A_{\alpha,\beta} = \{z \in \mathbb{C} : f_\alpha(z) = f_\beta(z)\}.$$

Since $f_\alpha \neq f_\beta$, $A_{\alpha,\beta}$ is countable. Hence,

$$A = \cup_{\alpha<\beta} A_{\alpha,\beta}$$

is of cardinality $\leq \aleph_1$. Under $\neg CH$, there is a $z_0 \in \mathbb{C} \setminus A$. Clearly, for all $\alpha < \beta < \omega_1$, $f_\alpha(z_0) \neq f_\beta(z_0)$ implying that $|\{f_\alpha(z_0) : \alpha < \omega_1\}| = \aleph_1$. Thus, (1) implies (2).

Conversely assume CH. Enumerate $\mathbb{C} = \{z_\alpha : \alpha < \omega_1\}$. Let D be a countable dense subset of $\mathbb{C}$. By transfinite induction we shall define distinct entire functions f_α, $\alpha < \omega_1$, such that whenever $\beta < \alpha$, $f_\alpha(z_\beta) \in D$. This will show that (2) implies (1).

Suppose $\alpha < \omega_1$ and entire functions f_β, $\beta < \alpha$, have been defined. We enumerate $\{f_\beta : \beta < \alpha\}$ as a sequence $\{g_n\}$ and $\{z_\beta : \beta < \alpha\}$ as $\{w_n\}$. We shall now construct an entire function f_α such that for every n,

$$f_\alpha(w_n) \in D \quad \& \quad f_\alpha(w_n) \neq g_n(w_n).$$

We shall choose a sequence $\{\epsilon_n\}$ of complex numbers and a complex number a so that the function

$$f_\alpha(z) = a + \sum_n \epsilon_n \Pi_{i=0}^n (z - w_i)$$

will satisfy all the requirements.

Choose any $a \in D \setminus \{g_0(w_0)\}$. Choose ϵ_0 such that $|\epsilon_0(z - w_0)| < 1/2^0$ for all $|z| < 1$ and $a + \epsilon_0(w_1 - w_0) \in D \setminus \{g_1(w_1)\}$. Such a ϵ_0 is possible to choose because

4.1 Analysis

$$\frac{(D \setminus \{g_1(w_1)\}) - a}{w_1 - w_0}$$

is dense in $\mathbb{C}$. Suppose, $\epsilon_0, \ldots, \epsilon_{n-1}$ have been chosen.

Choose ϵ_n such that

$$\forall |z| < n + 1 (|\epsilon_n(z - w_0) \cdots (z - w_n)| < 1/2^n)$$

and

$$a + \epsilon_0(w_{n+1} - w_0) + \cdots + \epsilon_n(w_{n+1} - w_0) \cdots (w_{n+1} - w_n) \in D \setminus \{g_{n+1}(w_{n+1})\}.$$

Since D is dense, such a ϵ_n is possible to choose.

Then the series

$$f_\alpha(z) = a + \sum_n \epsilon_n \Pi_{i=0}^n (z - w_i)$$

is uniformly convergent on each compact subsets of $\mathbb{C}$. Hence, f_α is entire. We have shown that (2) implies (1). ∎

We now proceed to prove Cantor's theorem on sets of uniqueness of a trigonometric series. As mentioned in the section on historical remarks in Chap. 2, the investigation of sets of uniqueness of a trigonometric series led Cantor to introduce well-ordered sets, ordinal numbers, and ultimately develop set theory.

Let $A \subset \mathbb{R}$ be a closed and bounded set and $\{A^{(\alpha)} : \alpha < \omega_1\}$ its Cantor–Bendixson derivatives. Since this a non-increasing transfinite sequence of closed subsets of $\mathbb{R}$, by Exercise 3.3.14, there is a $\alpha < \omega_1$ such that for all $\alpha < \beta < \omega_1$, $A^{(\alpha)} = A^{(\beta)}$. Let $\alpha_0 < \omega_1$ be the first such α. If $A^{(\alpha_0)} \neq \emptyset$, then by Proposition 4.1.8, $|A| \geq |A^{(\alpha_0)}| = \mathfrak{c}$. We have proved the following result.

Proposition 4.1.17 *Let $A \subset \mathbb{R}$ be compact. Then A is countable if and only if $A^{(\alpha)} = \emptyset$ for some $\alpha < \omega_1$.*

Assume now that $A \subset \mathbb{R}$ is countable and compact and α_0 is the first ordinal such that $A^{(\alpha_0)} = \emptyset$. By compactness of Cantor–Bendixson derivatives α_0 cannot be a limit ordinal, i.e., α_0 is a successor ordinal or 0. It is not hard to show that

$$A = \cup_{\alpha < \alpha_0}(A^{(\alpha)} \setminus A^{(\alpha+1)}).$$

Let

$$S \sim \sum_{n=-\infty}^{\infty} c_n e^{inx}$$

be a formal trigonometric series. We have defined sets of uniqueness of a trigonometric series S earlier in this note. Our goal now is to prove the following theorem of Cantor.

Theorem 4.1.18 *Every countable compact subset of $\mathbb{R}$ is a set of uniqueness of S.*

We remark that this result can be generalized for any countable closed subset of $\mathbb{R}$. We leave its proof as an exercise for the reader. We also remark that Cantor showed only that every closed subset of $\mathbb{R}$ whose Cantor–Bendixson derivative vanishes at a finite stage is a set of uniqueness of S. However, his arguments prove the above result mutatis mutandis.

We first state some results without proof that will be used to prove the above result of Cantor. For a function $f : (a, b) \to \mathbb{R}$, we define

$$\Delta^2 f(x+h) = f(x+h) + f(x-h) - 2f(x).$$

Lemma 4.1.19 (Schwarz) *Suppose f is continuous and*

$$D^2 f(x) = \lim_{h \to 0} \frac{\Delta^2 f(x, h)}{h^2}$$

exists for all x in an open interval (a, b). If $D^2 f \geq 0$ on (a, b), then f is convex. In particular, if $D^2 f = 0$ on (a, b), f is linear on (a, b).

We now fix a trigonometric series

$$S \sim \sum_{n=-\infty}^{\infty} c_n e^{inx}, \quad x \in \mathbb{R}.$$

For $x \in \mathbb{R}$, we say that the series S is convergent if

$$\lim_{N \to \infty} \sum_{n=-N}^{N} c_n e^{inx}$$

is convergent. Consider the following function:

$$f(x) = \frac{c_0}{2} x^2 - \sum_{n \neq 0} \frac{c_n}{n^2} e^{inx}, \quad x \in \mathbb{R}$$

obtained by formally integrating the series termwise twice. By Weierstrass M-test, f is continuous on $\mathbb{R}$ if the sequence $\{c_n\}_{n=-\infty}^{\infty}$ is bounded.

Lemma 4.1.20 (Riemann's First Lemma) *If the coefficients $\{c_n\}$ of S are bounded and $\sum_{n=-\infty}^{\infty} c_n e^{inx}$ converges to s for some real number x, then $D^2 f(x)$ exists and equals s.*

4.1 Analysis

Lemma 4.1.21 (Riemann's Second Lemma) *If $c_n \to 0$ as $|n| \to \infty$, then*

$$\lim_{h \to 0} \frac{\Delta^2 f(x, h)}{h} = 0$$

for every $x \in \mathbb{R}$.

Lemma 4.1.22 (Cantor) *If S converges to 0 for all x in a non-empty open interval, then $c_n \to 0$ as $|n| \to \infty$.*

We have now given all the required results to prove Cantor's theorem on sets of uniqueness. The main idea of the proof is given by the following lemma.

Proposition 4.1.23 *Let $\{c_n\}$ be bounded. Suppose there exist m and l such that for every $x \in \mathbb{R}$,*

$$f(x) = \frac{c_0}{2} x^2 - \sum_{n \neq 0} \frac{c_n}{n^2} e^{inx} = m \cdot x + l. \tag{$*$}$$

Then $c_n = 0$ for all n.

Proof By substituting $x = \pi$ and $x = -\pi$ in $(*)$ and subtracting we get $m = 0$. We now have

$$f(x) = \frac{c_0}{2} x^2 - \sum_{n \neq 0} \frac{c_n}{n^2} e^{inx} = l. \tag{$**$}$$

By substituting $x = 0$ and $x = 2\pi$ and subtracting, we get $c_0 = 0$. Thus, we have

$$\sum_{n \neq 0} \frac{c_n}{n^2} e^{inx} = -l. \tag{$***$}$$

Since the series $\sum_{n \neq 0} \frac{c_n}{n^2} e^{inx}$ is uniformly convergent on $\mathbb{R}$, interchanging the order of integration and summation is possible. Thus from $(***)$ for each $k \neq 0$, we get

$$2\pi \frac{c_k}{k^2} = \sum_{n \neq 0} \int_0^{2\pi} \frac{c_n}{n^2} e^{i(n-k)x} dx = -l \int_0^{2\pi} e^{-ikx} dx = 0.$$

It follows that $c_k = 0$. Our proof is complete. ∎

Corollary 4.1.24 *Suppose $\{c_n\}$ is bounded and $D^2 f(x) = 0$ for all real x, then for every n, $c_n = 0$.*

Proof of Cantor's theorem Let $A \subset \mathbb{R}$ be countable and compact and $\{A^{(\alpha)} : \alpha < \omega_1\}$ its Cantor–Bendixson derivatives. Suppose α_0 is the first ordinal $\alpha < \omega_1$ such that $A^{(\alpha)} = \emptyset$. By the last lemma, we need to show that

$$f(x) = \frac{c_0}{2}x^2 - \sum_{n \neq 0} \frac{c_n}{n^2} e^{inx}$$

is convergent for all real number x and is linear.

Case 1: $\alpha_0 = 0$. Then $A = A^{(0)} = \emptyset$. Hence, S converges to 0 for every real x. By Lemma 4.1.22, $c_n \to 0$ as $|n| \to \infty$. In particular, $\{c_n\}$ is bounded. Hence, by Weierstrass M-test,

$$f(x) = \frac{c_0}{2}x^2 - \sum_{n \neq 0} \frac{c_n}{n^2} e^{inx}$$

is convergent for every real number and continuous. By Riemann's first lemma (4.1.20) $D^2 f(x) = 0$. By Schwarz lemma (4.1.19), $f(x)$ is linear on $\mathbb{R}$.

Case 2: $\alpha_0 = 1$. In this case A is a finite set. Enumerate A as

$$a_0 < a_1 < \cdots < a_m.$$

By Cantor's lemma, $c_n \to 0$ as $|n| \to \infty$. Hence, $f(x)$ exists for all real x and is continuous. By Riemann's first lemma, $f(x)$ is linear on each of $(-\infty, a_0)$, (a_m, ∞) and (a_i, a_{i+1}), $i < m$. Take any a_i. By Riemann's second lemma (4.1.21),

$$0 = \lim_{h \to 0+} \left(\frac{f(a_i + h) - f(a_i)}{h} - \frac{f(a_i - h) - f(a_i)}{-h} \right).$$

This implies that the slopes of the linear maps representing f(x) on one of the above intervals to the left of a_i and to the right of a_i are equal. Since $f(x)$ is continuous on $\mathbb{R}$, it follows that $f(x)$ is linear on $\mathbb{R}$.

Case 3. $\alpha_0 > 1$. We have observed that α_0 is a successor ordinal, say $\alpha_0 = \beta_0 + 1$. Then $A^{(\beta_0)}$ is a finite set, say

$$a_1 < \cdots < a_m.$$

Set $a_0 = -\infty$ and $a_{m+1} = \infty$. Then S converges to 0 on (a_0, a_1). By Cantor's lemma it follows that $\{c_n\}$ is bounded. Hence, $f(x)$ exists for all real x and is continuous.

For $\alpha \leq \alpha_0$, set

$$V^\alpha = \mathbb{R} \setminus A^{(\alpha)}.$$

By transfinite induction, we shall show that $f(x)$ is linear in each open interval contained in $V^{(\alpha)}$. Since $V^{(\alpha_0)} = \mathbb{R}$, it will follow that $f(x)$ is linear on $\mathbb{R}$ which will complete our proof.

For $\alpha = 0$, S vanishes on each interval contained in $\mathbb{R} \setminus A$. Hence, as argued before, $f(x)$ is linear on each open interval contained in $V^{(0)}$.

Assuming that f is linear on each non-empty open interval contained in $V^{(\alpha)}$, we show that f is linear on each of the largest open interval contained in $V^{(\alpha+1)}$. Let $[m, M] \subset V^{(\alpha+1)}$ be a compact non-degenerate interval. Since $[m, M]$ is compact, let

$$m \leq a_1 < \cdots < a_k \leq M$$

be all the end points of largest open intervals contained in $V^{(\alpha)}$ which belong to $[m, M]$. By the induction hypothesis, f is linear on each of the open intervals (m, a_1), (a_k, M), and (a_i, a_{i+1}), $1 \leq i < k$. By Riemann's second lemma, slopes of each of these linear maps are equal. By the continuity of f, it follows that f is linear on $[m, M]$. Since $[m, M]$ was an arbitrary compact interval contained in I, it follows that f is linear on I.

Finally, assume that $\alpha < \alpha_0$ is a limit ordinal and our hypothesis holds for all $\beta < \alpha$. Let $[m, M]$ be a compact non-degenerate interval contained in $V^{(\alpha)}$. By the compactness of $[m, M]$, $[m, M]$ is contained in $V^{(\beta)}$ for some $\beta < \alpha$. Hence, by the induction hypothesis, f is linear on $[m, M]$. The proof is seen now. ∎

4.2 Topology

In this section, we give applications of set theory to topology. We begin by proving a basic theorem in topology known as Tychonoff's theorem on compact spaces.

A topological space X is called *compact* if every open cover of X has a finite subcover. We now give several equivalent conditions for a topological space to be compact.

Proposition 4.2.1 *Let X be a topological space. The following conditions are equivalent:*

1. *X is compact.*
2. *Let $\mathcal{C}$ be a family of closed sets in X with f.i.p. Then $\cap \mathcal{C} \neq \emptyset$.*
3. *Let $\mathcal{A}$ be a family of sets with f.i.p. Then $\cap \{\overline{A} : A \in \mathcal{A}\} \neq \emptyset$.*
4. *If $\mathcal{F}$ is a filter on X, then $\cap \{\overline{A} : A \in \mathcal{F}\} \neq \emptyset$.*
5. *If $\mathcal{U}$ is an ultrafilter on X, then $\cap \{\overline{A} : A \in \mathcal{U}\} \neq \emptyset$.*

Proof Let X be compact and $\mathcal{C}$ a family of closed sets in X with f.i.p. If possible, suppose $\cap \mathcal{C} = \emptyset$. Then $\mathcal{U} = \{X \setminus C : C \in \mathcal{C}\}$ is an open cover of X. By (1), $\mathcal{U}$ has a finite subcover, say $\{X \setminus C_1, \ldots, X \setminus C_n\}$. But then $\cap_{i=1}^{n} C_i = \emptyset$ contradicting that $\mathcal{C}$ has f.i.p. We have proved that (1) implies (2). Using De Morgan law we similarly prove that (2) implies (1). The entirely trivial proof is left to the reader as an exercise.

(2) and (3) are easily seen to be equivalent.

Now assume (2) and let $\mathcal{F}$ be a filter on X. Then $\{\overline{A} : A \in \mathcal{F}\}$ is a family of closed sets with f.i.p. By (2), this family has non-empty intersection. Thus (2) implies (4). Next assume (4) and let $\mathcal{C}$ be a family of closed sets in X with f.i.p. As we have observed earlier, $\mathcal{C}$ is contained in a filter, say $\mathcal{F}$. By (4),

$$\emptyset \neq \cap \{\overline{A} : A \in \mathcal{F}\} \subset \cap \mathcal{C}.$$

So, (2) and (4) are equivalent.

Since every ultrafilter is a filter, (4) clearly implies (5). Next assume (5) and let $\mathcal{F}$ be a filter on X. Using Zorn's lemma we showed that $\mathcal{F}$ is contained in an ultrafilter, say $\mathcal{U}$. By (5),

$$\emptyset \neq \cap\{\overline{A} : A \in \mathcal{U}\} \subset \cap\{\overline{A} : A \in \mathcal{F}\}.$$

Our proof is complete now. ∎

Lemma 4.2.2 *Let X be a compact Hausdorff space and $\mathcal{U}$ is an ultrafilter on X, then $\cap\{\overline{A} : A \in \mathcal{U}\}$ contains exactly one point.*

Proof We already know that $\cap\{\overline{A} : A \in \mathcal{U}\}$ is non-empty. If possible suppose $x \neq y \in \cap\{\overline{A} : A \in \mathcal{U}\}$. Since X is Hausdorff there is an open set U such that $x \in U$ and $y \notin \overline{U}$. Then $U \cap \overline{A} \neq \emptyset$ for every $A \in \mathcal{U}$. Since U is open, $U \cap A \neq \emptyset$ for every $A \in \mathcal{U}$. Since $\mathcal{U}$ is maximal, $U \in \mathcal{U}$ by Theorem 2.3.16. This implies that $y \notin \cap\{\overline{A} : A \in \mathcal{U}\}$. This contradiction proves our result. ∎

Let $\{X_i : i \in I\}$ be a family of topological spaces, $X = \times_{i \in I} X_i$ and $\pi_i : X \to X_i$, $i \in I$, the projection map from X to X_i. The *product topology* on X is the smallest topology on X making each π_i continuous. So, the product topology on X is the smallest topology containing $\cup_{i \in I}\{\pi_i^{-1}(U) : U \text{ open in } X_i\}$. The family

$$\mathcal{B} = \{\cap_{i \in F} \pi_i^{-1}(U_i) : F \subset I \text{ finite } \& \ U_i \text{ open in } X_i\}$$

is a base of the product topology on X. From rudimentary results in topology, it follows that if $A \subset X$ and $\alpha \in X$, then $\alpha \in \overline{A}$ if and only if $A \cap V \neq \emptyset$ for every $V \in \mathcal{B}$ containing α.

Theorem 4.2.3 (Tychonoff's Theorem) *Let $\{X_i : i \in I\}$ be a family of compact spaces. Then $X = \pi_{i \in I} X_i$ with product topology is compact.*

Proof Let $\mathcal{U}$ be an ultrafilter on X. By the last proposition, it is sufficient to prove that $\cap\{\overline{A} : A \in \mathcal{U}\} \neq \emptyset$.

Fix an $i \in I$ and consider $\mathcal{A}_i = \{\pi_i(A) : A \in \mathcal{U}\}$. Since $\mathcal{U}$ has f.i.p., $\mathcal{A}_i$ has f.i.p. Since X_i is compact, by the last proposition,

$$B_i = \cap\{\overline{B} : B \in \mathcal{A}_i\} \neq \emptyset.$$

By AC, there is a $\alpha \in \times_{i \in I} B_i \subset X$.

We shall show that for every $A \in \mathcal{U}, \alpha \in \overline{A}$. Fix a basic open set $U = \cap_{i \in F} \pi_i^{-1}(U_i)$ containing α, where $F \subset I$ is finite and U_i open in X_i for each $i \in F$. As remarked above, it is sufficient to show that for every $A \in \mathcal{U}$, $A \cap U \neq \emptyset$. This will follow if $U \in \mathcal{U}$. Since $\mathcal{U}$ is closed under finite intersections, it will be sufficient to show that $\pi_i^{-1}(U_i) \in \mathcal{U}$ for each $i \in F$. Fix a $i \in F$. Since $\mathcal{U}$ is a maximal filter, $\pi_i^{-1}(U_i) \in \mathcal{U}$ if for every $A \in \mathcal{U}$, $\pi_i^{-1}(U_i) \cap A \neq \emptyset$, i.e., $U_i \cap \pi_i(A) \neq \emptyset$. This is true because $\alpha(i) \in \overline{\pi_i(A)}$.

The proof of Tychonoff's theorem is complete. ∎

Remark 4.2.4 Tychonoff's theorem for compact Hausdorff spaces $\{X_i : i \in I\}$ can be proved using only the statement *every filter on a non-empty set is contained in a*

4.2 Topology

maximal filter. This is interesting because the statement *every filter on a non-empty set is contained in a maximal filter* is weaker than full AC.

To see our assertion, in the above proof for each $i \in I$, let $\mathcal{U}_i$ be an ultrafilter containing $\mathcal{A}_i$. By Lemma 4.2.2, there is a unique $\alpha \in X$ such that for every $i \in I$, $\alpha(i) \in \bigcap \{\overline{A} : A \in \mathcal{U}_i\}$. (We are not using AC here.) In particular, $\alpha \in \times_{i \in I} B_i \subset X$. Our assertion is easily seen now.

It is very interesting to note that Kelly [3] showed that Tychonoff's theorem implies AC. We close this section by showing this.

Proposition 4.2.5 (Kelly) *Tychonoff's theorem implies AC.*

Proof Let $\{A_i : i \in I\}$ be a family of non-empty sets. For reasons that will become clear very soon, for each $i \in I$, set $\infty_i = A_i$. Then, as remarked in the introduction, $\infty_i \notin A_i$. Set $X_i = A_i \cup \{\infty_i\}$, i.e., $X_i = s(A_i)$.

Fix $i \in I$. Set
$$\mathcal{B}_i = \{U \subset A_i : A_i \setminus U \text{ is finite}\} \cup \{\infty_i\}.$$

Since $\mathcal{B}_i$ is closed under finite intersections, it is a base of a topology on X_i, say $\mathcal{T}_i$. It is easy to check that $(X_i, \mathcal{T}_i)$ is compact. Hence, by Tychonoff's theorem, $X = \times_{i \in I} X_i$ with the product topology is compact.

Now consider $\mathcal{C} = \{\pi_i^{-1}(A_i) : i \in I\}$. Since A_i is closed in X_i for each $i \in I$, $\mathcal{C}$ is a family of closed sets in X. Take any finite $F \subset I$. Without using AC, we have shown that $\times_{i \in F} A_i \neq \emptyset$. Choose a $\beta \in \times_{i \in F} A_i$. Then α defined by

$$\alpha(j) = \begin{cases} \beta(j) \text{ if } j \in F \\ \infty_j \text{ if } j \in I \setminus F \end{cases}$$

belongs to $\bigcap_{i \in F} \pi_i^{-1}(A_i)$ showing that $\mathcal{C}$ has f.i.p. Since X is compact, it follows that $\bigcap \mathcal{C} = \times_{i \in I} A_i \neq \emptyset$. ∎

We begin by giving a characterization of normal spaces which is of importance in the study of paracompact spaces. Let X be a topological space and $\mathcal{A} = \{A_i : i \in I\}$ a family of subsets of X. We call $\mathcal{A}$ *point finite* if for every $x \in X$, the set $\{i \in I : x \in A_i\}$ is finite and *locally finite* if every $x \in X$ is contained in an open set U such that $\{i \in I : A_i \cap U \neq \emptyset\}$ is finite. Note that every locally finite family of subsets of X is point finite. We call $\mathcal{A}$ a *cover* of X if $X = \cup_i A_i$. If, moreover, each A_i is open in X, we call $\mathcal{A}$ an *open cover* of X.

Proposition 4.2.6 *Let X be a topological space. The following conditions are equivalent.*

1. *The topological space X is normal.*
2. *For every point-finite open cover $\mathcal{U} = \{U_i : i \in I\}$ of X, there is an open cover $\mathcal{V} = \{V_i : i \in I\}$ of X such that for every $i \in I$, $\overline{V_i} \subset U_i$, where $\overline{V_i}$ denotes the closure of V_i.*

Proof We first prove that the statement (2) implies that X is normal. Let C_1 and C_2 be two disjoint closed subsets of X. Then $\mathcal{U} = \{X \setminus C_1, X \setminus C_2\}$ is a point-finite open cover of X. Hence, by (2), there exist open sets V_1 and V_2 with union X such that $\overline{V_i} \subset U_i, i = 1, 2$. Take $W_i = X \setminus \overline{V_i}, i = 1, 2$. Then W_1 and W_2 are open. Since $\overline{V_1} \cup \overline{V_2} = X$, $W_1 \cap W_2 = \emptyset$. Further, $W_i \supset C_i, i = 1, 2$. Thus, X is normal.

Conversely, assume that X is normal and $\mathcal{U} = \{U_i : i \in I\}$ is a point-finite open cover of X. Set $\mathbb{P}$ to be the set of all $\mathcal{W} = \{W_i : i \in I\}$ such that $\cup_i W_i = X$, each W_i is open and $\forall i \in I(W_i = U_i \vee \overline{W_i} \subset U_i)$.

To motivate the proof, we make an observation first. Suppose $\mathcal{W} = \{W_i : i \in I\} \in \mathbb{P}$ and there exists a $i_0 \in I$ such that $\overline{W_{i_0}} \not\subset U_{i_0}$. Note that

$$C = X \setminus \cup_{i \neq i_0} W_i \subset W_{i_0} = U_{i_0}$$

with C closed and U_{i_0} open. Since X is normal, there is an open set W such that $C \subset W \subset \overline{W} \subset U_{i_0}$. Now consider $\mathcal{W}' = \{W'_i : i \in I\}$, where

$$W'_i = \begin{cases} W & \text{if } i = i_0, \\ W_i & \text{if } i \neq i_0. \end{cases}$$

Then $\mathcal{W}' \in \mathbb{P}$, $\overline{W'_{i_0}} \subset U_{i_0}$, and $W'_i = W_i$ whenever $i \neq i_0$.

Motivated by this observation, we define a partial order $\leq$ on $\mathbb{P}$ as follows: Let $\mathcal{W}_j = \{W^j_i : i \in I\} \in \mathbb{P}, j = 1, 2$. Define

$$\mathcal{W}_2 \leq \mathcal{W}_1 \Leftrightarrow \forall i \in I(W^2_i \neq U_i \Rightarrow W^1_i = W^2_i).$$

A careful examination of the above proof implies that our proof will be complete if we show that $\mathbb{P}$ has a maximal element. We shall now use Zorn's lemma to show the existence of a maximal element in $\mathbb{P}$. Since $\mathcal{U} \in \mathbb{P}$, $\mathbb{P} \neq \emptyset$.

Take a chain $\mathcal{C}$ in $\mathbb{P}$. For each $i \in I$, define

$$W^0_i = \cap_{\mathcal{W} \in \mathcal{C}} W_i,$$

where $\mathcal{W} = \{W_i : i \in I\}$. We now show that $\mathcal{W}_0 = \{W^0_i : i \in I\} \in \mathbb{P}$.

It is clear that for every $i \in I$, either $W^0_i = U_i$ or $\overline{W^0_i} \subset U_i$. It remains to check that $X = \cup_{i \in I} W^0_i$. Take any $x \in X$. Since $\mathcal{U}$ is point-finite, $\{i \in I : x \in U_i\}$ is finite, say $\{i_1, \ldots, i_k\}$. If there is a j, $1 \leq j \leq k$, such that $W^0_{i_j} = U_{i_j}$, then $x \in W^0_{i_j} \subset \cup_{i \in I} W^0_i$.

Next assume that for all $1 \leq j \leq k$, $W^0_{i_j} \neq U_{i_j}$. Then for each $1 \leq j \leq k$, there is a $C_j = \{A^j_i : i \in I\} \in \mathcal{C}$ such that $A^j_{i_j} \neq U_{i_j}$. Since $\mathcal{C}$ is a chain, there is a $C = \{A_i : i \in I\} \in \mathcal{C}$ such that $C_j \leq C$ for all $1 \leq j \leq k$. It follows that there is a $1 \leq j \leq k$ such that $x \in A_{i_j} = W^0_{i_j} \subset \cup_{i \in I} W^0_i$.

It is easy to see that $\mathcal{W}_0$ is an upper bound of $\mathcal{C}$. Therefore, by Zorn's lemma, $\mathbb{P}$ has a maximal element. ∎

4.2 Topology

Paracompact spaces play an important role in mathematics. Let X be a topological space and $\mathcal{U}$ an open cover of X. An open cover $\mathcal{V}$ is called a *refinement* of $\mathcal{U}$ if every $V \in \mathcal{V}$ is contained in some $U \in \mathcal{U}$. The topological space X is called *paracompact* if every open cover of $\mathcal{U}$ admits a locally finite open refinement.

Exercise 4.2.7 Let $\mathcal{C}$ be a locally finite family of closed subsets of a topological space X. Show that $C = \cup \mathcal{C}$ is closed in X.

Exercise 4.2.8 Show that every paracompact space X is normal.

We present below a result of A. H. Stone which easily implies that every metric space is paracompact. It uses WOP crucially.

A family $\mathcal{B}$ of subsets of X is called *discrete* if every $x \in X$ is contained in some open set U that has non-empty intersection with at most one $B \in \mathcal{B}$. A countable union of discrete families of subsets of X is called σ-discrete. Let (X, d) be a metric space, $x \in X$ and $A, B \subset X$. We set

$$d(x, A) = \inf\{d(x, y) : y \in A\}$$

and

$$d(A, B) = \inf\{d(x, y) : x \in A \ \& \ y \in B\}.$$

Proposition 4.2.9 (Stone [4]) *Every open cover $\mathcal{U}$ of a metric space (X, d) has a σ-discrete open refinement.*

Proof By WOP we can well order $\mathcal{U}$ and write $\mathcal{U} = \{U_\alpha : \alpha < \kappa\}$, where κ is the cardinality of $\mathcal{U}$. For each $n \geq 1$ and $\alpha < \kappa$, define

$$U_{\alpha,n} = \{x \in U_\alpha : d(x, X \setminus U_\alpha) \geq 2^{-n}\}.$$

Using the triangle inequality, we can check that

$$d(U_{\alpha,n}, X \setminus U_{\alpha,n+1}) \geq 2^{-n-1}.$$

For each $\alpha < \kappa$ and $n \geq 1$, define

$$U^*_{\alpha,n} = U_{\alpha,n} \setminus \cup_{\beta < \alpha} U_{\beta,n+1}.$$

Note that if $\beta < \alpha < \kappa$, then

$$U^*_{\alpha,n} \subset X \setminus U_{\beta,n+1}.$$

It is not hard to check that for $\beta < \alpha < \kappa$ and $n \geq 1$,

$$d(U^*_{\alpha,n}, U^*_{\beta,n}) \geq 2^{-n-1}.$$

Now define
$$U_{\alpha,n}^- = \{x \in X : d(x, U_{\alpha,n}) < 2^{-n-3}\}.$$

Then each $U_{\alpha,n}^-$ is open and
$$d(U_{\alpha,n}^-, U_{\beta,n}^-) \geq 2^{-n-2},$$

for all $\beta < \alpha < \kappa$ and $n \geq 1$. For each $n \geq 1$, set
$$\mathcal{B}_n = \{U_{\alpha,n}^- : \alpha < \kappa\}.$$

Then $\mathcal{B}_n$ is discrete and each element of $\mathcal{B}_n$ is open and contained in some set belonging to $\mathcal{U}$.

We now show that $X = \cup \mathcal{B}$. Take any $x \in X$. Let α be the first ordinal less than κ such that $x \in U_\alpha$. Then it is not hard to show that $x \in U_{\alpha,n}^-$ for some $n \geq 1$. Hence, $\mathcal{B} = \cup_n \mathcal{B}_n$ is a σ-discrete open refinement of $\mathcal{U}$. ∎

By a standard result in topology, "*a regular space X is paracompact if and only if every open cover of X has a σ-discrete open refinement.*" Since every metric space is regular, we have the following result.

Theorem 4.2.10 (Stone [4]) *Every metric space is paracompact.*

Next we study a topology on some sets of ordinals and give an example of a subspace of a normal space which is not normal. Recall that given a linearly ordered set $(X, \leq)$, we defined open intervals in X. We called unions of open intervals open in X. This indeed defines a topology on X, called the *order topology*. For $x < y \in X$, $[x, y]$ will stand for the set $\{z \in X : x \leq z \leq y\}$.

Proposition 4.2.11 *Let $\alpha < \beta$ be ordinal numbers. Then $X = [\alpha, \beta]$ with order topology is a compact, Hausdorff space.*

Proof We leave it for the reader to show that X is Hausdorff. Let $\mathcal{U}$ be an open cover of X. Define
$$A = \{\gamma \in ON : \alpha \leq \gamma \leq \beta \ \& \ [\alpha, \gamma] \subset \cup_{i=1}^k U_i\}$$

for some finitely many $U_1, \ldots, U_k \in \mathcal{U}$. We show that every $\gamma \in [\alpha, \beta]$ belongs to A.

Let $\alpha \leq \gamma \leq \beta$ be such that every $\delta < \gamma$ belongs to A. Suppose $\gamma = \delta + 1$ is a successor ordinal. Get $U_1, \ldots, U_k \in \mathcal{U}$ such that $[\alpha, \delta] \subset \cup_{i=1}^k U_i$. Since $\mathcal{U}$ covers X, there is a $U_{k+1} \in \mathcal{U}$ containing γ. It follows that $\gamma \in A$.

Next assume that γ is a limit ordinal. Get a $U \in \mathcal{U}$ containing γ. Since U is open, there is a $\delta < \gamma$ such that $(\delta, \gamma] \subset U$. By hypothesis there exist $U_1, \ldots, U_k \in \mathcal{U}$ such that $[\alpha, \delta] \subset \cup_{i=1}^k U_i$. Then, $\{U_1, \ldots, U_k, U\}$ is a finite subcover of $[\alpha, \gamma]$. Our result is easily seen now by transfinite induction. ∎

We now give an example of a non-normal subspace of a normal space.

4.2 Topology

Example 4.2.12 Let $Y = [0, \omega_1]$ and $Z = [0, \omega_0]$. By the above proposition Y and Z are compact, Hausdorff spaces. By Tychonoff theorem $X = Y \times Z$ is a compact, Hausdorff space. It is a standard result in topology that every compact, Hausdorff space is normal. Hence, X is normal.

Take $T = X \setminus \{(\omega_1, \omega_0)\}$. We show that T is not normal. Let $A = \{\omega_1\} \times [0, \omega_0)$ and $B = [0, \omega_1) \times \{\omega_0\}$. These are two disjoint closed sets in T. Take any open set $U \supset A$. Then for every $n < \omega_0$ there is a $\alpha_n < \omega_1$ such that $(\alpha_n, \omega_1] \times \{n\} \subset U$. Get a countable ordinal $\alpha > \alpha_n$ for every $n < \omega_0$. Then $\cup_n (\alpha, \omega_1] \times \{n\} \subset U$.

Now let $V \supset B$ be open. Then there is a $n < \omega_0$ such that $(\alpha + 1, n) \in U \cap V$. It follows that T is not a normal subspace of X.

The space T in the above example is also known as *Tychonoff plank*. We give one more example of a non-normal subspace of a normal space. This example does not use ordinal numbers so may be accessible to a wider class of readers. However, the argument is completely set theoretic.

Example 4.2.13 Let A be any uncountable set and B a countably infinite subset. Let $C = A \setminus B$. Fix any $p \in \{0, 1\}^A$. Set

$$X = \{0, 1\}^A \setminus \{p\}.$$

We show that X is not a normal subspace of the normal space $\{0, 1\}^A$.

Define

$$F_1 = \{f \in X : f \equiv p \text{ on } B\}$$

and

$$F_2 = \{f \in X : f \equiv p \text{ on } C\}.$$

The sets F_1 and F_2 are disjoint and closed in X. Take any open set U in X containing F_2. We show that
$$F_1 \cap \overline{U} \neq \emptyset.$$

Our claim will follow.

For each finite $F \subset B$, choose an $p_F \in F_2$ such that $p_F \equiv p$ on F. Such a $p_F \in X$ exists because B is infinite. Since $p_F \in U$, there is a basic neighborhood of p_F, say W_F, contained in U. Without any loss of generality, we can assume that

$$W_F = \{q \in X : q \equiv p \text{ on } I_F \cup J_F \text{ and } q \equiv 1 - p \text{ on } K_F\},$$

for some finite subsets I_F, K_F of B and some finite subset J_F of C. Further, we can assume that $F \subset I_F$. Consider the function

$$p_0(a) = \begin{cases} p(a) & \text{if } a \in B \cup \cup_F J_F, \\ 1 - p(a) & \text{if } a \in C \setminus \cup_F J_F. \end{cases}$$

Since C is uncountable and $\cup_F J_F$ countable, $p_0 \neq p$. Clearly, $p_0 \in F_1$. We show that $p_0 \in \text{cl}(U)$. Take any basic neighborhood W of p_0. Get finite subsets L, M of A such that
$$W = \{q \in X : q \equiv p \text{ on } L \text{ and } q \equiv 1 - p \text{ on } M\}.$$

Set $F = L \cap B$. Then $W_F \cap W \neq \emptyset$.

To see this first observe the following two facts:

(a) $p_0 \equiv p$ on $I_F \cup J_F$ (by definition) and $p_0 \equiv 1 - p$ on M (because $p_0 \in W$). Thus, $M \cap (I_F \cup J_F) = \emptyset$.

(b) As $L \cap B = F \subset I_F$, $K_F \subset B$ and $K_F \cap I_F = \emptyset$, $L \cap K_F = \emptyset$.

It follows that $(I_F \cup J_F \cup L) \cap (M \cup K_F) = \emptyset$. Since these sets are countable and A uncountable there is a $q \in X$ satisfying

$$q(a) = \begin{cases} p(a) & \text{if } a \in I_F \cup J_F \cup L, \\ 1 - p(a) & \text{if } a \in M \cup K_F. \end{cases}$$

Such a $q \in W_F \cap W$.

We close this section by giving examples of two normal spaces whose product is not normal.

Example 4.2.14 Let $X = [0, \omega_1)$. We now prove that X is normal. Let C_1 and C_2 be two disjoint closed sets in X. If possible, suppose both are unbounded. Then as they are CUBs, as observed earlier, $C_1 \cap C_2 \neq \emptyset$. Since this is not the case, at least one of C_1, C_2 is bounded. Without any loss of generality assume that C_1 is bounded. Take $\alpha < \omega_1$ such that $C_1 \subset [0, \alpha]$. As shown above, $[0, \alpha]$ is compact and Hausdorff, so normal. Therefore, there exist disjoint open sets U_1 and U_2 in $[0, \alpha]$ such that $C_1 \subset U_1$ and $C_2 \cap [0, \alpha] \subset U_2$. Since $[0, \alpha] = [0, \alpha + 1)$, U_1 and U_2 are open in X also. Note that U_1 and $U_2 \cup (\alpha, \omega_1)$ are disjoint open sets in $[0, \omega_1)$ such that $U_1 \supset C_1$ and $U_2 \cup (\alpha, \omega) \supset C_2$. Thus we have shown that X is normal.

Example 4.2.15 Let $Y = [0, \omega_1)$ and $X = Y \times [0, \omega_1]$. We now show that X is not normal. Consider $C_1 = Y \times \{\omega_1\}$ and $C_2 = \{(\alpha, \alpha) \in X : \alpha < \omega_1\}$. Then C_1 and C_2 are disjoint closed sets in X. Let $U \supset C_1$ be open. To prove that X is not normal, we shall show that $C_2 \cap \overline{U} \neq \emptyset$. Start with a $(\alpha_0, \omega_1) \in C_1$. Then there exists a $\alpha_1 > \alpha_0$ such that $\{\alpha_0\} \times [\alpha_1, \omega_1] \subset U$. In particular, $(\alpha_0, \alpha_1) \in U$. Suppose $\alpha_0 < \alpha_1 < \cdots < \alpha_n < \omega_1$ have been defined so that for all $i < n$, $(\alpha_i, \alpha_{i+1}) \in U$. Now choose $\alpha_{n+1} > \alpha_n$ so that $\{\alpha_n\} \times [\alpha_{n+1}, \omega_1] \subset U$. Proceeding in this way we get an increasing sequence $\alpha_0 < \alpha_1 < \alpha_2 < \alpha_3 < \ldots < \omega_1$ such that for each n, $(\alpha_n, \alpha_{n+1}) \in U$. Let $\alpha = \sup_n \alpha_n < \omega_1$. Then $(\alpha, \alpha) \in C_1 \cap \overline{U}$.

4.3 Linear Algebra

Let V be a vector space over a field k and $\emptyset \neq W \subset V$. We call W a (*vector*) *subspace* of V if for every $x, y \in W$ and $\lambda \in k$, $x - y, \lambda x \in W$. Since $W \neq \emptyset$, $0 \in W$. For a subset E of V, the *span* of E, denoted by span(E), is the smallest vector subspace of V containing E. It equals

$$\operatorname{span}(E) = \{\sum_{i=1}^{n} \lambda_i x_i \in V : \lambda_1, \ldots, \lambda_n \in k \ \& \ x_1, \ldots, x_n \in E\}.$$

A subset A of V is called *independent* if whenever $\lambda_1, \ldots, \lambda_n \in k$, $x_1, \ldots, x_n \in A$ and $\sum_{i=1}^{n} \lambda_i x_i = 0$, each $\lambda_i = 0$. If $x = \sum_{i=1}^{n} \lambda_i x_i$, we will say that x is a *linear sum* of $x_1, \ldots, x_n$. Note the following facts:

1. $\emptyset$ is independent.
2. $A \subset V$ is independent if and only if every finite subset of A is independent.
3. If A is independent and $y \in V \setminus \operatorname{span}(A)$, then $A \cup \{y\}$ is independent: Suppose $\lambda y + \sum_{i=1}^{n} \lambda_i x_i = 0$, $x_1, \ldots, x_n \in A$. If possible, suppose $\lambda \neq 0$. Then $y = \sum_{i=1}^{n} -\lambda^{-1} \lambda_i x_i$. Hence, $y \in \operatorname{span}(A)$ which contradicts our assumption.
4. If W is a subspace of V and $a \in V \setminus W$, then

$$\operatorname{span}(W \cup \{a\}) = \{x + \lambda a : x \in W \ \& \ \lambda \in k\}.$$

A subset B of V is called a *basis* of V if it is independent and $\operatorname{span}(B) = V$. The trivial proof of the following lemma is omitted.

Lemma 4.3.1 *Let V be a vector space over a field k and $B \subset V$. Then the following two statements are equivalent.*

1. *B is a basis of V.*
2. *Every $x \in V$ can be expressed uniquely as a linear sum $x = \sum_{i=1}^{n} \lambda_i x_i$, where $\lambda_1, \ldots, \lambda_n \in k$ and $x_1, \ldots, x_n \in B$.*

Proposition 4.3.2 *Let V be a vector space over a field k and A an independent subset of V. Then there is a basis B of V containing A.*

Proof Define
$$\mathbb{P} = \{C \subset V : C \supset A \ \& \ C \text{ is independent}\}.$$

Since $A \in \mathbb{P}$, $\mathbb{P} \neq \emptyset$. Partially order $\mathbb{P}$ by inclusion $\subset$. It is easy to check that every chain in $\mathbb{P}$ has an upper bound in $\mathbb{P}$. Hence, $\mathbb{P}$ has a maximal element, say B. By fact (3) stated above, every maximal independent set is a basis of V. Thus, B is a basis of V containing A. ∎

Corollary 4.3.3 *Every vector space over a field has a basis.*

Proposition 4.3.4 *Let V be a vector space over a field k and B_1, B_2 two basis of V. Suppose B_1 is infinite. Then B_2 is infinite and $|B_1| = |B_2|$.*

Proof If possible, suppose B_2 is finite. Then there is a finite set $A \subset B_1$ such that $\text{span}(A) \supset B_2$. This implies that $\text{span}(A) = \text{span}(B_2) = V$. Since B_1 is infinite, there is an $a \in B_1 \setminus A$. In particular, $a \in \text{span}(A)$ contradicting that B_1 is independent.

Since B_2 is independent and spans V, for every finite subset A of B_1 there is a unique finite $B \subset B_2$ such that $\text{span}(A) = \text{span}(B)$. This defines a function f from the set C_1 of all finite subsets of B_1 into the set C_2 of all finite subsets of B_2 such that for every $A \in C_1$, $\text{span}(A) = \text{span}(f(A))$. Similarly, we have a map g from C_2 into C_1 such that for every $B \in C_2$, $\text{span}(B) = \text{span}(g(B))$. Clearly, $g = f^{-1}$. Since B_1 and B_2 are infinite, we have

$$|B_1| = |C_1| = |C_2| = |B_2|.$$

∎

Corollary 4.3.5 *The cardinality of any two basis of a vector space over a field is equal.*

If the vector space has an infinite basis, then the corollary is proved above. We advise the reader to read the proof in case of V having a finite basis from any book in algebra or linear algebra. The cardinality of a basis of V is called the *dimension* of V.

Let V_1 and V_2 be vector spaces over k. A map $\Lambda : V_1 \to V_2$ is called *linear* if for every $x, y \in V_1$ and $\lambda, \mu \in k$,

$$\Lambda(\lambda x + \mu y) = \lambda \Lambda(x) + \mu \Lambda(y).$$

$\Lambda : V_1 \to V_2$ is called an *isomorphism* if Λ is a bijection and linear. In this case, $\Lambda^{-1} : V_2 \to V_1$ is also linear. V_1 and V_2 are called *isomorphic* if there is an isomorphism from V_1 onto V_2.

Exercise 4.3.6 Let V_1 and V_2 be vector spaces over k. Show the following.

1. Let B be a basis of V_1 and $\Lambda_1, \Lambda_2 : V_1 \to V_2$ linear. Then

$$\Lambda_1 = \Lambda_2 \Leftrightarrow \Lambda_1 | B = \Lambda_2 | B.$$

2. Every map $f : B \to V_2$ has a unique linear extension $\Lambda : V_1 \to V_2$, where B is a basis of V_1.
3. V_1 and V_2 are isomorphic if and only if their dimensions are equal.

4.3 Linear Algebra

Let V be a vector space over a field k. A linear map f from V to k is called a *linear functional* on V. The set of all linear functionals on V is denoted by V^* and is called the *dual* of V. For $\Lambda_1, \Lambda_2 \in V^*$ and $\alpha \in k$, we define

$$(\Lambda_1 + \Lambda_2)(x) = \Lambda_1(x) + \Lambda_2(x), \quad x \in V$$

and

$$(\alpha \Lambda_1)(x) = \alpha \cdot \Lambda_1(x), \quad x \in V.$$

It is easy to check that these make V^* a vector space over k.

Exercise 4.3.7 If the dimension of V is λ, show that the dimension of V^* is $|k|^\lambda$.

Exercise 4.3.8 Let V be an infinite dimensional vector space over a field k. Show that the dimension of its dual V^* is $\geq \mathfrak{c}$. Conclude that if is a vector space over an field k of dimension $\aleph_0$, then there is no vector space W whose dual is V.

(**Hint:**) First note that it is sufficient to prove the result for $k = \mathbb{Z}_2$. Now use the fact that there is a family $\mathcal{A}$ of infinite subsets of $\mathbb{N}$ of cardinality $\mathfrak{c}$ such for any two $A \neq B \in \mathcal{A}$, $A \cap B$ is finite. (Exercise 2.1.39.)

The set of all real numbers is a vector space over the field $\mathbb{Q}$. Any basis of the vector space $\mathbb{R}$ over $\mathbb{Q}$ (which exists as shown above) is called a *Hamel basis* of $\mathbb{R}$.

Proposition 4.3.9 *1. Every Hamel basis is uncountable.*
2. The cardinality of a Hamel basis equals $\mathfrak{c}$.
3. Let V_1 and V_2 be two uncountable vector spaces over $\mathbb{Q}$. Then V_1 and V_2 are isomorphic if and only if $|V_1| = |V_2|$.

Proof (1). Let $B \subset \mathbb{R}$ be independent over $\mathbb{Q}$. Then every element $x \in \text{span}(B)$ has a unique linear representation $\sum_{i=1}^n q_i x_i$, where $q_1, \ldots, q_n \in \mathbb{Q}$ and $x_1, \ldots, x_n \in B$. Suppose B is countable. Then $\text{span}(B)$ is countable. Hence, B is not a Hamel basis.

(2) Let B be a Hamel basis. Then $\mathfrak{c} = \aleph_0^{<\aleph_0} \cdot |B|^{<\aleph_0} = \aleph_0 \cdot |B| = |B|$.

(3) Let V be an uncountable vector space over $\mathbb{Q}$ and B a basis of V. By the arguments contained in the proofs of (1) and (2), we see that $|V| = |B|$. Therefore, dimensions of V_1 and V_2 are equal if and only if $|V_1| = |V_2|$. (3) follows. ∎

Proposition 4.3.10 *Let V_1 and V_2 be two vector spaces over a field k. Suppose W is a vector subspace of V_1, $a \in V_1 \setminus W$ and $f : W \to V_2$ a linear map. Then we have the following:*

1. There is a linear map $g : \text{span}(W \cup \{a\}) \to V_2$ extending f.
2. There is a linear map $F : V_1 \to V_2$ extending f.

Proof (1) First recall that

$$\text{span}(W \cup \{a\}) = \{x + \lambda a : x \in W \ \& \ \lambda \in k\}.$$

Take any $b \in V_2$ and define

$$g(x + \lambda a) = f(x) + \lambda b, \quad x \in W, \lambda \in k.$$

It is fairly routine to check that g is well defined, linear, and it extends f.

(2) Consider the following poset $\mathbb{P}$,

$\{(V, h) : V_1 \supset V \supset W, V \text{ a subspace of } V_1, h : V \to V_2 \text{ linear extending } f\}$.

Since $(W, f) \in \mathbb{P}, \mathbb{P} \neq \emptyset$.

For $(W_1, h_1), (W_2, h_2) \in \mathbb{P}$, define

$$(W_1, h_1) \leq (W_2, h_2) \Leftrightarrow W_2 \supset W_1 \ \& \ h_2|W_1 = h_1.$$

It can be easily checked that $\leq$ is a partial order on $\mathbb{P}$ with every chain in $\mathbb{P}$ bounded above. By Zorn's lemma, $\mathbb{P}$ has a maximal element, say (V_0, F). Since (V_0, F) is maximal, by (1), $V_0 = V_1$. This completes the proof of (2). ∎

This is the main technique used in proving the very useful Hahn–Banach theorem in functional analysis.

Theorem 4.3.11 *Let* $(X, ||\cdot||)$ *be a normed linear space over* $\mathbb{R}$, Y *a subspace of* X *and* $f : X \to \mathbb{R}$ *a continuous linear functional. Then there is a continuous linear functional* $F : X \to \mathbb{R}$ *extending* f *such that* $||F|| = ||f||$.

The proof proceeds in two steps as above. In step (1), it is possible to choose $b \in \mathbb{R}$ in such a way that $||g|| = ||f||$. To see this, let W be a subspace of X, $a \in X \setminus W$ and $f : W \to \mathbb{R}$ a continuous linear functional.

Take arbitrary $x, y \in W$. Then

$$|f(x) + f(y)| = |f(x+y)| \leq ||f|| \cdot ||x+y|| = ||f|| \cdot ||(x-a) + (y+a)||.$$

In particular,
$$f(x) + f(y) \leq ||f|| \cdot (||x-a|| + ||y+a||).$$

This implies that

$$f(x) - ||f|| \cdot ||x-a|| \leq ||f|| \cdot ||y+a|| - f(y).$$

We choose any $b \in \mathbb{R}$ such that for every $x, y \in W$,

$$f(x) - ||f|| \cdot ||x-a|| \leq b \leq ||f|| \cdot ||y+a|| - f(y).$$

We set $g(a) = b$ and for any $\lambda \in \mathbb{R}$ and any $x \in W$, we put

$$g(x + \lambda a) = f(x) + \lambda b.$$

This works.

Then using Zorn's lemma as in (2) by taking in $\mathbb{P}$ only the norm preserving extensions of f, the proof is completed. ∎

Remark 4.3.12 We see that AC implies the Hahn–Banach theorem. However, there are class models of $ZF + \neg AC$ in which the Hahn–Banach theorem holds. Thus, the Hahn–Banach theorem is weaker than AC (in ZF).

4.4 Algebra

Vector spaces over $\mathbb{Q}$ can be identified with a class of interesting groups. Let $(G, +)$ be an abelian group. For $a \in G$ and m a positive integer, let ma denote the element of G obtained by adding a to itself m times. A non-zero element x of G is called a *torsion element* of G if there is a positive integer n such that $nx = 0$. We call G *torsion free* if it contains no torsion element. The group G is called *divisible* if for every $0 \neq x \in G$ and every positive integer n, there is a $y \in G$ such that $ny = x$. If, moreover, G is torsion free, such a y is unique. We then write $y = \frac{x}{n}$. For every rational number $q = \frac{m}{n}, n > 0$, we define $q \cdot x = \frac{mx}{n}$. The following result is quite easy to see now.

Proposition 4.4.1 *Let $(G, +)$ be a group. Then G is a torsion-free divisible abelian group if and only if there is a map $(q, x) \to q \cdot x$ from $\mathbb{Q} \times G \to G$ which makes G a vector space over $\mathbb{Q}$.*

Corollary 4.4.2 *Let G_1 and G_2 be two uncountable, torsion-free, divisible abelian groups. Then G_1 and G_2 are isomorphic as groups if and only if $|G_1| = |G_2|$.*

Example 4.4.3 Let $1 \leq n < m$ be positive integers. Then $\mathbb{R}^n$ and $\mathbb{R}^m$ with canonical group structures are isomorphic.

As a direct application of Zorn's lemma, a result of fundamental importance is proved in commutative algebra. Let R be a commutative ring with identity 1. An *ideal* I in R is a subring of R such that whenever $x \in I$ for every $y \in R$, $y \cdot x \in I$. If $I \neq R$, I is called a *proper ideal*. Note that an ideal I is a proper ideal if and only if $1 \notin I$. A proper ideal I is called a *maximal ideal* if it is a maximal element of the set of all proper ideals with respect to inclusion $\subset$. Maximal ideals play a very fundamental role in commutative algebra.

Proposition 4.4.4 *Every proper ideal in a commutative ring R with identity is contained in a maximal ideal.*

Proof Let I be a proper ideal in R. Consider

$$\mathbb{P} = \{J \subset R : 1 \notin J, J \text{ is an ideal in } R \text{ and } J \supset I\}.$$

Since $I \in \mathbb{P}$, $\mathbb{P} \neq \emptyset$. Partially order $\mathbb{P}$ with respect to the inclusion $\subset$. If $\mathcal{C}$ is a chain in $\mathbb{P}$, then $J = \cup \mathcal{C}$ is an ideal not containing 1 which contains every ideal belonging to $\mathcal{C}$. Hence, J is an upper bound of $\mathcal{C}$. The result now follows from Zorn's lemma. ∎

Example 4.4.5 Let $\{R_a : a \in A\}$ be an infinite family of commutative rings with identity. Consider $R = \times_{a \in A} R_a$. It is a commutative ring with identity with addition and multiplication defined coordinatewise with 0 taken to be $a \to 0_a$ and $1, a \to 1_a$. R is called the *direct product* of $\{R_a : a \in A\}$. Define

$$I = \{x \in R : \exists \text{ finite } B \subset A \forall a \in A \setminus B(x(a) = 0)\}.$$

Since A is infinite, I is a proper ideal of R. I is called the *direct sum* of $\{R_a : a \in A\}$ and is denoted by $\oplus_{a \in A} R_a$. By the above proposition, I is contained in a maximal ideal of R.

Let $\{\mathbb{F}_a : a \in A\}$ be a family of fields and $\mathcal{U}$ an ultrafilter on A. The ultraproduct $(\times_{a \in A} \mathbb{F}_a)/\mathcal{U}$ of $\{\mathbb{F}_a : a \in A\}$ will be denoted by $\mathbb{F}^{\mathcal{U}}$. Let $[\alpha], [\beta] \in \mathbb{F}^{\mathcal{U}}$. We define

$$[\alpha] + [\beta] = [\alpha + \beta] \ \& \ [\alpha] \cdot [\beta] = [\alpha \cdot \beta],$$

where $(\alpha + \beta)(a) = \alpha(a) + \beta(a)$ and $(\alpha \cdot \beta)(a) = \alpha(a) \cdot \beta(a)$, $a \in A$. Note that $[\alpha] + [\beta]$ and $[\alpha] \cdot [\beta]$ are well defined. We define

$$0 = [\overline{0}] \ \& \ 1 = [\overline{1}],$$

where $\overline{0}(a) = 0_a$ and $\overline{1}(a) = 1_a$, 0_a and 1_a being the additive and multiplicative identity of $\mathbb{F}_a$, respectively, $a \in A$.

Exercise 4.4.6 Show the following.

1. $\mathbb{F}^{\mathcal{U}}$ is a field.
2. Characteristic of $\mathbb{F}^{\mathcal{U}}$ is p, p a prime, if and only if

$$\{a \in A : \text{characteristic}(\mathbb{F}_a) = p\} \in \mathcal{U}.$$

3. If $\{a \in A : \text{characteristic}(\mathbb{F}_a) = 0\} \in \mathcal{U}$, then characteristic$(\mathbb{F}^{\mathcal{U}}) = 0$.
4. Let P denote the set of all prime numbers and $\mathcal{U}$ a free ultrafilter on P. For each $p \in P$, let $\mathbb{F}_p$ denote the finite field of cardinality p. Show that characteristic$(\mathbb{F}^{\mathcal{U}}) = 0$.
5. Let P denote the set of all primes and $\mathcal{U}$ a free ultrafilter on P. For each $p \in P$, let $\mathbb{F}_p$ denote the field with p elements. Then characteristic$(\mathbb{F}^{\mathcal{U}}) = 0$.

4.4 Algebra

(**Hint:** For (2), assume that the characteristic of $\mathbb{F}^{\mathcal{U}}$ is p, p a prime. This implies that $p = 0$ in $\mathbb{F}^{\mathcal{U}}$. Therefore, by definition,

$$\{a \in A : p = 0 \text{ in } \mathbb{F}_a\} \in \mathcal{U}.$$

For (3), fix any $n > 1$. Note that $n\overline{1} = \overline{(n1_a)}$. By our hypothesis, $\{a \in A : n1_a \neq 0\} \in \mathcal{U}$. If $n\overline{1} = 0$, then $\{a \in A : n1_a \neq 0\} \in \mathcal{U}$. Then $\mathcal{U}$ contains two disjoint sets which is a contradiction.

For (4), for every $n > 1$, $A = \{q \in P : q > n\}$ is cofinite. Since $\mathcal{U}$ is a free ultrafilter, $A \in \mathcal{U}$. Hence, $\{p \in P : n\overline{1} = 0\} \subset P \setminus A$. It follows that $n\overline{1} \neq = 0$.

We are now going to prove that every field has an algebraic closure. Let $\mathbb{F}_0$ be a field and A the set of all finite, non-empty subsets of $\mathbb{F}_0[X]$, the set of all polynomials over $\mathbb{F}_0$ in a single variable X. We know that for each finite $a \in A$, there is a field extension $\mathbb{F}_a$ of $\mathbb{F}_0$ in which each polynomial $f(X) \in a$ has a root.

For each $f \in \mathbb{F}_0[X]$, let $B_f = \{a \in A : f \in a\}$. If $f_1, \ldots, f_n \in \mathbb{F}_0[X]$, $\{f_1, \ldots, f_n\} \in \cap_{i=1}^n B_{f_i}$. Thus, $\{B_f : f \in \mathbb{F}_0[X]\}$ has finite intersection property. Hence, there is an ultrafilter $\mathcal{U}$ containing each B_f.

Set $\mathbb{F}_1 = \mathbb{F}^{\mathcal{U}}$, the ultraproduct of the family $\{\mathbb{F}_a : a \in A\}$ of fields. As noted in the above exercise $\mathbb{F}^{\mathcal{U}}$ is a field. It is not hard to show that there is an embedding of $\mathbb{F}_0$ in $\mathbb{F}^{\mathcal{U}}$.

Now take any $f \in \mathbb{F}_0[X]$ and consider $B_f \in \mathcal{U}$. For each $a \in B_f$, let $z_a \in \mathbb{F}_a$ be a root of f. Set

$$z(a) = \begin{cases} z_a & \text{if } a \in B_f, \\ 0 & \text{if } a \notin B_f. \end{cases}$$

It is easy to check that $[z]$ is a root of f in $\mathbb{F}_1$. Thus, we have proved the following result.

Proposition 4.4.7 *Every field $\mathbb{F}_0$ has an extension $\mathbb{F}_1$ such that every $f \in \mathbb{F}_0[X]$ has a root in $\mathbb{F}_1$.*

Corollary 4.4.8 *Every field $\mathbb{F}_0$ is a subfield of an algebraically closed field $\mathbb{F}_\infty$.*

Proof Using the last proposition repeatedly we get a sequence of fields

$$\mathbb{F}_0 \subset \mathbb{F}_1 \subset \mathbb{F}_2, \ldots$$

such that for every n, $\mathbb{F}_{n+1}$ is an extension of $\mathbb{F}_n$ containing a root of each polynomial $f(X) \in \mathbb{F}_n[X]$. Now take $\mathbb{F}_\infty = \cup_n \mathbb{F}_n$. ∎

Let $\overline{\mathbb{F}_0}$ be the intersection of all algebraically closed subfields of $\mathbb{F}_\infty$ containing $\mathbb{F}_0$. We see that $\overline{\mathbb{F}_0}$ is the smallest algebraically closed field containing $\mathbb{F}_0$. The field $\overline{\mathbb{F}_0}$ is called the *algebraic closure* of $\mathbb{F}_0$.

We shall now state a series of facts. They are checked by routine cardinal arithmetic and their proofs are left for the reader.

(I) Let G be a group, $A \subset G$ and H the subgroup of G generated by A. Then H is countable if A is countable. If A is uncountable, then $|H| = |A|$.

In the remaining facts k is either F_p, p prime, or $\mathbb{Q}$ and K an extension field of k.

(II) Let L be a subfield of K (containing k). Let $X = L[X_1, \ldots, X_n]$, the ring of polynomials over L in n-variables, or $X = L(X_1, \ldots, X_n)$, the field of rational functions over L in n variables. If L is countable, so is X. If L is uncountable, then $|X| = |L|$.

Let L be a subfield of K. An element $\alpha \in K$ is called *algebraic over L* if there is a non-zero polynomial $f(X) \in L[X]$ such that $f(\alpha) = 0$. Elements of K which are not algebraic over L are called *transcendental* over L.

(III) The set of all elements of K algebraic over a countable subfield L is countable. Suppose L is an uncountable subfield of K. Then the set of all elements of K algebraic over L is of cardinality $|L|$.

(IV) For $A \subset K$ and L a subfield of K, let $L[A]$ denote the smallest subring of K containing $L \cup A$ and $L(A)$ the smallest subfield of K containing $L \cup A$. Then

$$L[A] = \cup_n \{f(a_1, \ldots, a_n) : f \in L[X_1, \ldots, X_n] \ \& \ a_1, \ldots, a_n \in A\}$$

and $L(A)$ equals the union over all $n \geq 1$ 9f

$$\{\frac{f(a_1, \ldots, a_n)}{g(a_1, \ldots, a_n)} : f, g \in L[X_1, \ldots, X_n] \ \& \ a_1, \ldots, a_n \in A \ \& \ g(a_1, \ldots, a_n) \neq 0\}.$$

If A is countable, so are $k[A]$ and $k(A)$. If A is uncountable, then

$$|k[A]| = |A| = |k(A)|.$$

Let L be a subfield of K and $a_1, \ldots, a_n \in K$. We say that $a_1, \ldots, a_n$ are *algebraically independent* over L if there is no non-zero polynomial $f \in L[X_1, \ldots, X_n]$ such that $f(a_1, \ldots, a_n) = 0$.

Let $a_1, \ldots, a_n$ be algebraically independent over L. Then no a_i, $1 \leq i \leq n$, is algebraic over L: Let $g(X_i) \in L[X_i]$ be a non-zero polynomial over L such that $g(a_i) = 0$. Then $f(a_1, \ldots, a_n) = 0$, where $f(X_1, \ldots, X_n) = g(X_i)$. This contradicts that $a_1, \ldots, a_n$ are algebraically independent over L.

A subset A of K is called *algebraically independent* over L if each finite subset of A is algebraically independent over L.

Proposition 4.4.9 *Let $A \subset K$ be algebraically independent over a subfield L of K and $b \in K$. Then $A \cup \{b\}$ is algebraically independent over L if and only if b is not algebraic over $L(A)$.*

Proof Suppose b is algebraic over $L(A)$. Then there is a non-zero polynomial $f(X) \in L(A)[X]$ such that $f(b) = 0$. From the description of $L(A)$ given in fact (IV), it is easy to see that $A \cup \{b\}$ is not algebraically independent.

For the converse, assume that $A \cup \{b\}$ is not algebraically independent over L. Since A is algebraically independent over L, there is a polynomial $f \in$

$L[X_1, \ldots, X_n, Y]$ with the coefficient of at least one monomial containing a positive power of Y non-zero and $a_1, \ldots, a_n \in A$ such that $f(a_1, \ldots, a_n, b) = 0$. This implies that b is algebraic over $L(A)$. ∎

A direct application of Zorn's lemma shows the following.

Proposition 4.4.10 *Let K be a field extension of k, L a subfield of K and $A \subset K$ algebraically independent over L. Then A is contained in a maximal algebraically independent subset B over L. Moreover, every element of K is algebraic over $L(B)$.*

Let A be a maximal algebraically independent subset of K over L. Then every element $x \in L(A)$ has a unique representation

$$x = \frac{f(a_1, \ldots, a_n)}{g(a_1, \ldots, a_n)},$$

where $f, g \in L[X_1, \ldots, X_n], a_1, \ldots, a_n \in A$ and $g(a_1, \ldots, a_n) \neq 0$. Using this fact, it is not hard to show that for every field F and every map $f : A \to F$, there is a unique homomorphism $F : L(A) \to F$ such that $F|A = f$.

For the remaining part of this section, *we assume that K is an uncountable algebraically closed extension of k.* A maximal algebraically independent subset of K over k is called a *transcendence basis* of K. Then by facts stated above, we see that

$$|A| = |k(A)| = |K|.$$

The second equality holds because K is algebraic over $k(A)$. This in turn implies that any two transcendence bases of K are of the same cardinality.

Remark 4.4.11 Arguing as in the proof of "the cardinality of any two basis of a vector space are equal," we can show that if K is an algebraically closed extension of k of any cardinality, then the cardinality of any two transcendence bases of K is equal. We leave the proof when K is countable as an exercise for the reader. The cardinality of a transcendence basis of K is called the *transcendence degree* of K.

Theorem 4.4.12 *Let K_1 and K_2 are two algebraically closed fields. Then K_1 and K_2 are isomorphic if and only if K_1 and K_2 are of the same characteristic and same transcendence degree.*

Proof Only if part being trivial, we prove only the if part of the result. Since K_1 and K_2 are of the same characteristic their prime fields are the same, say k. Fix transcendence basis A_1 and A_2 of K_1 and K_2. Since $|A_1| = |A_2|$, there is a bijection $f : A_1 \to A_2$. Then as observed above there exist homomorphisms $F : k(A_1) \to k(A_2)$ and $G : k(A_2) \to k(A_1)$ of f and f^{-1}, respectively. By uniqueness of extension stated above, $G = F^{-1}$, i.e., $F : k(A_1) \to k(A_2)$ is an isomorphism. Since K_1 and K_2 are algebraically closed and algebraic extensions of $k(A_1)$ and $k(A_2)$, respectively. By standard arguments from algebra, F can be extended to an isomorphism $H : K_1 \to K_2$. ∎

Theorem 4.4.13 *Let K_1 and K_2 be two uncountable algebraically closed fields of the same characteristic. Then K_1 and K_2 are isomorphic if and only if $|K_1| = |K_2|$.*

Proof We only need to prove if part. Since K_1 and K_2 are uncountable and $|K_1| = |K_2|$, the transcendence degrees of K_1 and K_2 are equal. Our result follows directly from Theorem 4.4.12. ∎

4.5 Measure Theory

Set theory is very useful in measure theory and this can't be exaggerated. In this section, we shall restrict ourselves to some of the results in measure theory requiring the set theory presented in this note. Interested readers may see [1] for more delicate results.

Let X be a non-empty set and $\mathcal{A}$ a family of subsets of X. We define

$$\neg \mathcal{A} = \{A \subset X : X \setminus A \in \mathcal{A}\},$$

$$\mathcal{A}_s = \{\cup_{i=1}^n A_i : A_1, \ldots, A_n \in \mathcal{A}, n = 1, 2, \ldots\},$$

$$\mathcal{A}_d = \{\cap_{i=1}^n A_i : A_1, \ldots, A_n \in \mathcal{A}, n = 1, 2, \ldots\},$$

$$\mathcal{A}_\sigma = \{\cup_{i=1}^\infty A_i : A_1, A_2, \ldots \in \mathcal{A}\}$$

and

$$\mathcal{A}_\delta = \{\cap_{i=1}^\infty A_i : A_1, A_2, \ldots \in \mathcal{A}\}.$$

We say that $\mathcal{A}$ is *closed under complementations* if $\neg \mathcal{A} = \mathcal{A}$; it is *closed under finite unions (intersections)* if $\mathcal{A}_s = \mathcal{A}$ (resp. $\mathcal{A}_d = \mathcal{A}$) and *closed under countable unions (intersections)* if $\mathcal{A}_\sigma = \mathcal{A}$ (resp. $\mathcal{A}_\delta = \mathcal{A}$).

Exercise 4.5.1 $(\mathcal{A}_s)_s = \mathcal{A}_s$, $(\mathcal{A}_d)_d = \mathcal{A}_d$, $(\mathcal{A}_\sigma)_\sigma = \mathcal{A}_\sigma$ and $(\mathcal{A}_\delta)_\delta = \mathcal{A}_\delta$.

A family $\mathcal{F}$ of subsets of a non-empty set X is called a *field* of subsets of X if

1. $\emptyset \in \mathcal{F}$;
2. $\mathcal{F}$ is closed under complementations, i.e., whenever $A \in \mathcal{F}$, $X \setminus A \in \mathcal{F}$; and
3. $\mathcal{F}$ is closed under finite unions, i.e., whenever $A, B \in \mathcal{F}$, $A \cup B \in \mathcal{F}$.

Thus, a field $\mathcal{F}$ on X is a family of subsets which is closed under Boolean operation complementations and finite unions and contains $\emptyset$ and X.

Example 4.5.2 Let $\mathcal{F} = \{A \subset \mathbb{N} : A$ or $X \setminus A$ is finite $\}$. Show that $\mathcal{F}$ is a field on $\mathbb{N}$.

Exercise 4.5.3 Show that a field $\mathcal{F}$ is closed under finite intersections and $A \setminus B, A \triangle B \in \mathcal{F}$ whenever $A, B \in \mathcal{F}$.

4.5 Measure Theory

Let X be a non-empty set. Then its power set $\mathcal{P}(X)$ is a field. It is also clear that the intersection of any non-empty family of fields is a field. Now let $\mathcal{G} \subset \mathcal{P}(X)$. Then the intersection of all fields of subsets of X containing $\mathcal{G}$ is a field on X. It is the smallest field on X containing $\mathcal{G}$. We shall denote it by $\mathcal{F}(\mathcal{G})$. In this case, we say that $\mathcal{F}(\mathcal{G})$ is the *field generated by* $\mathcal{G}$ or that $\mathcal{G}$ is a *generator* of $\mathcal{F}(\mathcal{G})$. A field is called *countably generated* if it has a countable generator. Our first result follows from one of the general lemmas we proved toward the end of Chap. 3. Still we shall give a proof to make the result look simple and transparent.

Proposition 4.5.4 *Every countably generated field is countable.*

Proof Let X be a non-empty set and $\mathcal{G} \subset \mathcal{P}(X)$ countable.
Inductively define
$$\mathcal{G}_0 = \mathcal{G} \cup \neg \mathcal{G} \cup \{\emptyset\}.$$

If $n+1$ is odd, define
$$\mathcal{G}_{n+1} = (\mathcal{G}_n)_s.$$

If $n+1$ is even, then
$$\mathcal{G}_{n+1} = \mathcal{G}_n \cup \neg \mathcal{G}_n.$$

By induction on n we easily see that each $\mathcal{G}_n$ is countable and contained in $\mathcal{F}(\mathcal{G})$. Also $\cup_n \mathcal{G}_n$ is a field implying that it equals $\mathcal{F}(\mathcal{G})$. Since each $\mathcal{G}_n$ is countable, so is $\cup_n \mathcal{G}_n = \mathcal{F}(\mathcal{G})$. ∎

A family $\mathcal{A}$ of a subsets of a non-empty set X is called a σ-*field* on X if it satisfies the following conditions:

1. $\emptyset \in \mathcal{A}$;
2. $\mathcal{A}$ is closed under complementations; and
3. $\mathcal{A}$ is closed under countable unions, i.e., whenever $A_1, A_2, A_3, \ldots \in \mathcal{A}$, $\cup_n A_n \in \mathcal{A}$.

Exercise 4.5.5 Let $\mathcal{A}$ be a σ-field on X and $A_1, A_2, A_3, \ldots \in \mathcal{A}$. Show that $\cap_n A_n$, $\limsup_n A_n$ and $\liminf_n A_n$ are in $\mathcal{A}$.

$\mathcal{P}(X)$ is a σ-field on X and the intersection of a non-empty family of σ-fields on X is a σ-field on X. Hence, as before, given any $\mathcal{G} \subset \mathcal{P}(X)$ there is a smallest σ-field, denoted by $\sigma(\mathcal{G})$, on X containing $\mathcal{G}$. We call $\sigma(\mathcal{G})$ the σ-*field generated by* $\mathcal{G}$, and also we call $\mathcal{G}$ a *generator* of $\sigma(\mathcal{G})$.

Theorem 4.5.6 *Every countably generated σ-field is of cardinality $\leq \mathfrak{c}$.*

Proof Let X be a non-empty set and $\mathcal{G} \subset \mathcal{P}(X)$ countable. By transfinite induction, we define a transfinite sequence $\mathcal{G}_\alpha$, $\alpha < \omega_1$, of families of subsets of X as follows:

$$\mathcal{G}_0 = \mathcal{G} \cup \neg\mathcal{G} \cup \{\emptyset\}.$$

Suppose $\alpha < \omega_1$ and $\mathcal{G}_\beta$, $\beta < \alpha$, have been defined.

Consider the case when $\alpha = \beta + 1$ is a successor ordinal. If α is odd, define

$$\mathcal{G}_\alpha = (\mathcal{G}_\beta)_\sigma.$$

If $\alpha = \beta + 1$ is even, define

$$\mathcal{G}_\alpha = \mathcal{G}_\beta \cup \neg\mathcal{G}_\beta.$$

In case α is a limit ordinal, we define

$$\mathcal{G}_\alpha = \cup_{\beta<\alpha} \mathcal{G}_\beta.$$

By transfinite induction, it is easy to see that for each $\alpha < \omega_1$,

$$\mathcal{G}_\alpha \subset \sigma(\mathcal{G}).$$

Hence,

$$\cup_{\alpha<\omega_1} \mathcal{G}_\alpha \subset \sigma(\mathcal{G}).$$

$\cup_{\alpha<\omega_1} \mathcal{G}_\alpha$ is clearly closed under complementations. We now show that $\cup_{\alpha<\omega_1} \mathcal{G}_\alpha$ is closed under countable unions. Let

$$A_1, A_2, A_3, \ldots \in \cup_{\alpha<\omega_1} \mathcal{G}_\alpha.$$

For each n get a $\alpha_n < \omega_1$ such that $A_n \in \mathcal{G}_{\alpha_n}$. Get an even ordinal $\beta < \omega_1$ such that each $\alpha_n < \beta$. Then $\cup_n A_n \in \mathcal{G}_{\beta+1}$. It follows that

$$\cup_{\alpha<\omega_1} \mathcal{G}_\alpha = \sigma(\mathcal{G}).$$

We know that

$$\aleph_0^{\aleph_0} = \mathfrak{c}^{\aleph_0} = \mathfrak{c}.$$

Using these cardinal identities and standard facts on cardinal arithmetic, by transfinite induction, it is fairly routine to check that

$$\forall \alpha < \omega_1 (|\mathcal{G}_\alpha| \leq \mathfrak{c}).$$

4.5 Measure Theory

Hence,
$$|\sigma(\mathcal{G})| = |\cup_{\alpha<\omega_1} \mathcal{G}_\alpha| \leq \aleph_1 \cdot \mathfrak{c} = \mathfrak{c}.$$

∎

Let $\mathcal{A}$ be a σ-field on X and $\emptyset \neq A \in \mathcal{A}$. We call A an *atom* of $\mathcal{A}$ if there does not exist $\emptyset \neq B \in \mathcal{A}$ which is properly contained in A. Note that a $\emptyset \neq A \in \mathcal{A}$ is an atom of A if and only if for every $B \in \mathcal{A}$ either $B \cap A = \emptyset$ or $A \subset B$. This is because if a $B \in \mathcal{A}$ does not satisfy any of these two conditions then $B \cap A$ is a non-empty set in $\mathcal{A}$ which is properly contained in A.

Proposition 4.5.7 *Let $\mathcal{A} = \sigma(\mathcal{G})$ be a σ-field on X and $\emptyset \neq A \in \mathcal{A}$. Then A is an atom of $\mathcal{A}$ if and only if for every $B \in \mathcal{G}$*

$$\text{either } B \cap A = \emptyset \text{ or } A \subset B. \tag{$*$}$$

Proof Set
$$\mathcal{B} = \{B \in \mathcal{A} : B \text{ satisfies } (*)\}.$$

By our hypothesis, $\mathcal{G} \subset \mathcal{B}$. $\emptyset \in \mathcal{B}$. If $B \in \mathcal{B}$, then clearly $X \setminus B \in \mathcal{B}$ implying $\mathcal{B}$ is closed under complementations. Now assume that $B_1, B_2, B_3, \ldots \in \mathcal{B}$. If every $B_n \cap A = \emptyset$, then $A \cap \cup_n B_n = \emptyset$. If there is a B_n such that $B_n \cap A \neq \emptyset$, then $A \subset B_n \subset \cup_m B_m$. It follows that $\mathcal{B} \subset \mathcal{A} = \sigma(\mathcal{G})$ is a σ-field containing $\mathcal{G}$. Hence, $\mathcal{B} = \sigma(\mathcal{G})$ completing the proof of our lemma. ∎

We call a σ-field $\mathcal{A}$ on X *atomic* if X is the union of all atoms of $\mathcal{A}$. In this case every $B \in \mathcal{A}$ is the union of all the atoms of $\mathcal{A}$ contained in B. For $A \subset X$ and $\epsilon = 0$ or 1, we define

$$A^\epsilon = \begin{cases} A & \text{if } \epsilon = 0, \\ X \setminus A & \text{if } \epsilon = 1. \end{cases}$$

We now assume that $\mathcal{A}$ is countably generated and fix a sequence $\{A_n\}$ of sets in $\mathcal{A}$ that generate $\mathcal{A}$. For any $\bar{\epsilon} = (\epsilon_0, \epsilon_1, \epsilon_2, \ldots) \in \{0, 1\}^\mathbb{N}$, we define

$$A(\bar{\epsilon}) = \cap_{n=0}^\infty A_n^{\epsilon_n}.$$

Then for every $\bar{\epsilon}$ and any A_n in the generator of $\mathcal{A}$, either $A_n \cap A(\bar{\epsilon}) = \emptyset$ or $A_n \supset A(\bar{\epsilon})$. Hence, by Proposition 4.5.7, each non-empty $A(\bar{\epsilon})$ is an atom of $\mathcal{A}$.

Now take any $x \in X$. For a n, set $\epsilon_n = 0$ is $x \in A_n$ else put $\epsilon_n = 1$. Then $x \in A(\bar{\epsilon})$. We have now proved the following result:

Theorem 4.5.8 *Every countably generated σ-field is atomic.*

Using this we get a surprising result.

Theorem 4.5.9 *Every σ-field is either finite or of cardinality $\geq \mathfrak{c}$.*

Proof Let $\mathcal{A}$ be an infinite σ-field on a set X. Choose a sequence $\{A_n\}$ of distinct elements of $\mathcal{A}$. Consider $\sigma(\{A_n\})$. Since it is countably generated, by the last proposition, it is atomic. The number of its atoms cannot be finite because $\sigma(\{A_n\})$ is infinite. Let $C_1, C_2, C_3, \ldots$ be a sequence of distinct atoms of $\sigma(\{A_n\})$. All possible subunions of C_n's are in $\mathcal{A}$ and distinct. Hence, $|\mathcal{A}| \geq \mathfrak{c}$. ∎

Let $X = \mathbb{R}^n$. The σ-field generated by the family of all open sets in $\mathbb{R}^n$ is called the *Borel σ-field* of $\mathbb{R}^n$. It is denoted by $\mathcal{B}_{\mathbb{R}^n}$ and its members are called *Borel sets*. Consider the collection

$$\mathcal{G} = \{\times_{i=1}^n (a_i, b_i) : a_1, \ldots, a_n, b_1, \ldots, b_n \in \mathbb{Q}\}.$$

Then $\mathcal{G}$ is countable and all its elements are open in $\mathbb{R}^n$. Further, every open set is a union of sets in $\mathcal{G}$. Hence, $\mathcal{G}$ is a countable generator of $\mathcal{B}_{\mathbb{R}^n}$. $\mathcal{B}_{\mathbb{R}^n}$ being countably generated, $|\mathcal{B}_{\mathbb{R}^n}| \leq \mathfrak{c}$.

Every open rectangle $\times_{i=1}^n (a_i, b_i)$, $a_1, \ldots, a_n, b_1, \ldots, b_n \in \mathbb{R}$, is open. Hence, $|\mathcal{B}_{\mathbb{R}^n}| \geq \mathfrak{c}$. Now using Theorem 4.5.6 we have proved the following interesting result.

Theorem 4.5.10 $|\mathcal{B}_{\mathbb{R}^n}| = \mathfrak{c}$.

Lebesgue defined a σ-field $\mathcal{L}$ on $\mathbb{R}$ containing all open intervals and an extended non-negative real-valued function $\lambda : \mathcal{L} \to [0, \infty]$ satisfying the following properties.

1. $\lambda(\emptyset) = 0$.
2. (Countable Additive) For every sequence $\{A_n\}$ of pairwise disjoint sets in $\mathcal{L}$,

$$\lambda(\cup_n A_n) = \sum_n \lambda(A_n).$$

3. $\lambda((a, b)) = b - a$ for every $-\infty < a < b < \infty$.
4. (Translation Invariant) Whenever $A \in \mathcal{L}$, $A + x \in \mathcal{L}$ and $\lambda(A + x) = \lambda(A)$, $x \in \mathbb{R}$.
5. (Complete) Whenever $A \in \mathcal{L}$ and $\lambda(A) = 0$, for every $B \subset A (B \in \mathcal{L})$.

The σ-field $\mathcal{L}$ is called the *Lebesgue σ-field* on $\mathbb{R}$ and λ the *Lebesgue measure*. Sets in $\mathcal{L}$ are called *Lebesgue measurable*. Since $\mathcal{L}$ contains all open intervals, $\mathcal{L} \supset \mathcal{B}_{\mathbb{R}}$, i.e., all Borel sets in $\mathbb{R}$ are Lebesgue measurable. The reader may see [5] for the construction of the Lebesgue σ-field and the Lebesgue measure. The significance of these lie in the fact that we get a very large family of subsets of $\mathbb{R}$ where the notion of length can be extended.

Any extended non-negative real-valued set function defined on a σ-field satisfying conditions (1) and (2) is called a *measure*. If, moreover, it also satisfies condition (5), it is called a *complete measure*.

4.5 Measure Theory

For $s \in \{0, 1\}^{<\mathbb{N}}$ and $\epsilon = 0, 1$, $|s|$ will denote the length of s and $s \, \epsilon$ the concatenation of s and ϵ. e will denote the empty sequence of 0s and 1s. For $\alpha \in \{0, 1\}^{\mathbb{N}}$ and $n \in \mathbb{N}$, $\alpha|n = (\alpha(0), \ldots, \alpha(n-1))$.

In the following example $s, t \in \{0, 1\}^{<\mathbb{N}}$ and $\epsilon = 0, 1$.

Example 4.5.11 Define a system $\{I_s : s \in \{0, 1\}^{<\mathbb{N}}\}$ of non-empty closed subintervals of $[0, 1]$ satisfying the following conditions:

1. $I_e = [0, 1]$.
2. $(|s| = |t| > 0 \wedge s \neq t) \Rightarrow I_s \cap I_t = \emptyset$.
3. If $I_s = [a_s, b_s]$, then $I_{s\,0} = [a_s, a_s + \frac{b_s - a_s}{3}]$ and $I_{s\,1} = [b_s - \frac{b_s - a_s}{3}, b_s]$. Thus, $I_{s\,0}$ and $I_{s\,1}$ are obtained from I_s as follows: remove middle open $\frac{1}{3}$rd subinterval of I_s from I_s. The remaining part is a disjoint union of two closed subintervals of I_s. The left subinterval is $I_{s\,0}$ and the right one $I_{s\,1}$.

For $n \geq 0$, set
$$C_n = \cup_{|s|=n} I_s.$$

Then
 (a) Each C_n is compact and non-empty.
 (b) For all $n \geq 0$, $C_n \supset C_{n+1}$.
 (c) For all n, $\lambda(C_n) = (2/3)^n$.

We set
$$C = \cap_n C_n.$$

The set C is called *Cantor ternary set*. By property (c) above, $\lambda(C) = 0$. Hence, by the completeness of λ, every subset of C is Lebesgue measurable.

Take any $\alpha \in \{0, 1\}^{\mathbb{N}}$. By Cantor intersection theorem [1, Proposition 2.1.29], $\cap_n I_{\alpha|n}$ is a singleton, say $f(\alpha)$. If $\alpha \neq \beta$, then there is a positive integer n such that $\alpha|n \neq \beta|n$. This implies that $f(\alpha) \neq f(\beta)$. Thus, we have defined a one-to-one map $f : \{0, 1\}^{\mathbb{N}} \to C$. It now follows that $|C| = \mathfrak{c}$.

Exercise 4.5.12 Show that $f : \{0, 1\}^{\mathbb{N}} \to C$ is a homeomorphism.

Since λ is a complete measure, we now have the following result.

Proposition 4.5.13 $|\mathcal{L}| = 2^{\mathfrak{c}}$. *Hence, there is a Lebesgue measurable set which is not Borel.*

Remark 4.5.14 The above proposition only shows that there is a Lebesgue measurable set which is not Borel but does not give any example of such a set. See [1] for examples of Lebesgue measurable sets which are not Borel.

Using AC, we now give an example of a non-Lebesgue measurable set.

Example 4.5.15 For $x, y \in \mathbb{R}$, define

$$x \sim y \Leftrightarrow x - y \in \mathbb{Q}.$$

Then $\sim$ is an equivalence relation on $\mathbb{R}$. For $x \in \mathbb{R}$, let $[x] = x + \mathbb{Q}$ denote the equivalence class of x. Since $[x]$ is dense in $\mathbb{R}$, $[x] \cap (0, 1) \neq \emptyset$ for every $x \in \mathbb{R}$. By AC, there is a set S of reals such that for every $x \in \mathbb{R}$, $|S \cap [x] \cap (0, 1)| = 1$. Note that

$$(0, 1) \subset \cup \{x + r : x \in S \ \& \ r \in \mathbb{Q} \cap (-1, 1)\} \subset (0, 2).$$

We claim that S is not Lebesgue measurable. Suppose not. Enumerate $\mathbb{Q} \cap (-1, 1) = \{r_n : n = 0, 1, 2, \ldots\}$ and set $A_n = S + r_n$. If possible, suppose for some $n \neq m$, $A_n \cap A_m \neq \emptyset$. Then there exist $x, y \in S$ such that $x + r_n = y + r_m$. Then $x - y = r_m - r_n \neq 0$. This implies that $x \sim y$. This is a contradiction because $x \neq y \in S$ implies $x \nsim y$. Thus, A_n's are pairwise disjoint. By the translation invariance of the Lebesgue measure for every n, $\lambda(A_n) = \lambda(S)$.

For each N, $\cup_{n=0}^{N} A_n \subset (0, 2)$. Therefore,

$$\sum_{n=0}^{N} \lambda(A_n) = N\lambda(S) \leq 2.$$

Hence, $\lambda(S) = 0$. By countable additivity of λ, $1 = \lambda((0, 1)) \leq \sum_n \lambda(A_n) = 0$. This contradiction proves that S is not Lebesgue measurable.

Remark 4.5.16 The above example also shows that there is no extended real-valued translation invariant measure μ on $\mathcal{P}(\mathbb{R})$ such that the μ measure of every bounded set is finite.

The next result was proved by Banach. However, its proof is beyond the scope of this proof.

An extended non-negative, real-valued set function $\mu : \mathcal{A} \to [0, \infty]$, where $\mathcal{A}$ is a σ-field on a non-empty set Ω is called a *finitely additive measure* if it satisfies the following two conditions:

1. $\mu(\emptyset) = 0$.
2. (Finitely Additive) For pairwise disjoint $A_1, \ldots, A_n \in \mathcal{A}$,

$$\mu(\cup_1^n A_i) = \sum_1^n \mu(A_i).$$

Theorem 4.5.17 (Banach) *Let $n = 1, 2$. Then there exists a finitely additive measure μ on $\mathcal{P}(\mathbb{R}^n)$ such that $\mu([0, 1]^n) = 1$ and whenever A and B are congruent subsets of $\mathbb{R}^n$, $\mu(A) = \mu(B)$.*

4.5 Measure Theory

Hausdorff showed that Banach's theorem is false for $n > 2$. This will be proved in the next chapter.

Exercise 4.5.18 Let $\mathcal{F}$ be a field of subsets of a non-empty set X and $\mu : \mathcal{F} \to [0, \infty]$ a finitely additive measure. Show the following:

1. Whenever $A \subset B$ are sets in $\mathcal{F}$, $\mu(A) \leq \mu(B)$.
2. For every $A_1, \ldots, A_n \in \mathcal{F}$,

$$\mu(\cup_{i=1}^n A_i) \leq \sum_{i=1}^n \mu(A_n).$$

Proposition 4.5.19 *There is a finitely additive probability measure P on* $(\mathbb{N}, \mathcal{P}(\mathbb{N}))$ *such that for every* $n \in \mathbb{N}$, $P(\{n\}) = 0$.

Proof Let $\mathcal{U}$ be a free ultrafilter on $\mathbb{N}$. Define P on $\mathcal{P}(\mathbb{N})$ by

$$P(A) = \begin{cases} 1 & \text{if } A \in \mathcal{U}, \\ 0 & \text{if } A \notin \mathcal{U}. \end{cases}$$

The probability so defined has all the desired properties. ∎

Remark 4.5.20 Since on every infinite set X there is a free ultrafilter $\mathcal{U}$, the above proposition holds for all infinite sets X.

The next result, due to P. S. Alexandrov, is the first deep and beautiful result on Borel sets. It states that every uncountable Borel subset of $\mathbb{R}$ is of cardinality c. It is of historical importance too. First, it says that CH holds for Borel sets. Second the technique used to prove it was completely new. It gave rise to a new operation on sets, initially called operation($\mathcal{A}$), presumably $\mathcal{A}$ standing for Alexandrov. However, Luzin (considered to be the father of descriptive set theory) later named it Souslin operation after his favorite student. Then Souslin was a young and exciting mathematician who spotted an error in a proof of the great French mathematician Lebesgue. Sets obtained by operation($\mathcal{A}$) on a system of Borel sets were called A-sets and later analytic sets. Souslin proved the following beautiful result: *A set* $B \subset \mathbb{R}$ *is Borel if and only if both A and* $\mathbb{R} \setminus A$ *are analytic* leading to very deep results on Borel sets. Thus, it would be appropriate to consider Alexandrov's theorem the first and most instrumental result that gave birth to descriptive set theory which is a beautiful, deep, and useful branch of mathematics. Interested readers may see [1] to learn more on Borel sets and its applications.

Equip $\mathbb{N}$ with discrete topology and $\Sigma = \mathbb{N}^\mathbb{N}$ with the product topology. For a finite sequence $(n_0, \ldots, n_{k-1})$ of natural numbers, we define

$$\Sigma(n_0, \ldots, n_{k-1}) = \{\alpha \in \Sigma : \forall i < k(\alpha(i) = n_i)\}.$$

The collection
$$\{\Sigma(n_0, \ldots, n_{k-1}) : (n_0, \ldots, n_{k-1}) \in \mathbb{N}^{<\mathbb{N}}\}$$
is countable and a base for the topology of Σ. Now consider the so-called discrete metric d on $\mathbb{N}$ defined by
$$d(m, n) = \begin{cases} 1 & \text{if } m \neq n, \\ 0 & \text{if } m = n. \end{cases}$$

It is a complete metric on $\mathbb{N}$ inducing the discrete topology. Hence, their countable product Σ is a complete metric space. Define
$$\rho(\alpha, \beta) = \begin{cases} \frac{1}{\min\{i:\alpha(i)\neq\beta(i)\}+1} & \text{if } \alpha \neq \beta, \\ 0 & \text{if } \alpha = \beta, \end{cases}$$
where $\alpha, \beta \in \Sigma$. Then ρ is a complete metric on Σ inducing the product topology. We need to state a simple fact before proving our next result on Borel sets.

Lemma 4.5.21 *Let X be a topological space with a countable base. Then we can write $X = Y \cup Z$ where Z is countable and open and for every open set U in X, $U \cap Y \neq \emptyset \Rightarrow U \cap Y$ is uncountable.*

Proof Let $\{U_n\}$ be a countable base for the topology of X and take $Z = \cup_n \{U_n : U_n \text{ is countable}\}$ and $Y = X \setminus Z$. ∎

We point out that in the above lemma either Y or Z can be empty. However, if X is uncountable then Y is an uncountable closed set in X with no isolated points.

We shall use the following result on Borel sets. Its proof uses mainly topology and is somewhat involved. Interested readers may see [1] for the proof.

Proposition 4.5.22 *Every Borel subset of $\mathbb{R}^n$ is a continuous image of Σ.*

Theorem 4.5.23 (Alexandrov) *Every uncountable Borel subset of $\mathbb{R}^n$ is of cardinality $\mathfrak{c}$.*

Proof Let $B \subset \mathbb{R}^n$ be an uncountable Borel set. Recall that we have given an example of a complete metric ρ on Σ inducing its topology. We assume that $\mathbb{R}^n$ has the usual distance metric which is complete on $\mathbb{R}^n$. If (X, d) is a metric space and $A \subset X$, we define
$$\text{diameter}(A) = \sup\{d(x, y) : x, y \in A\}.$$

By the last proposition, there is a continuous surjection $f : \Sigma \to B$. Let $X \subset \Sigma$ be such that $|X \cap f^{-1}(x)| = 1$ for every $x \in B$. Since B is uncountable, X is uncountable. By the last lemma, write $X = Y \cup Z$ where Z is countable and open in X and for every open set U in X, $U \cap Y \neq \emptyset \Rightarrow U \cap Y$ is uncountable. Since X is uncountable and Z countable, Y is uncountable with no isolated points. Further, $f|Y$ is one to one.

4.5 Measure Theory

We shall define families of open sets $\{U_s : s \in \{0, 1\}^{<\mathbb{N}}\}$ and $\{V_s : s \in \{0, 1\}^{<\mathbb{N}}\}$ in Σ and $\mathbb{R}^n$, respectively, satisfying the following conditions. In the following conditions $s, t \in \{0, 1\}^{<\mathbb{N}}$ and $\epsilon = 0, 1$.

1. $U_e = \Sigma$ and $V_e = \mathbb{R}^n$.
2. $(|s| = |t| > 0 \wedge s \neq t) \Rightarrow (\overline{U_s} \cap \overline{U_t} = \emptyset = \overline{V_s} \cap \overline{V_t})$.
3. If $|s| > 0$, then diameter$(U_s) < 2^{-|s|}$ and diameter$(V_s) < 2^{-|s|}$.
4. $\overline{U_{s^\wedge \epsilon}} \subset U_s \wedge \overline{V_{s^\wedge \epsilon}} \subset V_s$.
5. $U_s \cap Y \neq \emptyset$.
6. $f(U_s) \subset V_s$.

Assuming that $\{U_s : s \in \{0, 1\}^{<\mathbb{N}}\}$ and $\{V_s : s \in \{0, 1\}^{<\mathbb{N}}\}$ satisfying the above conditions have been defined, we complete the proof first. Let $\alpha \in \{0, 1\}^{\mathbb{N}}$. Then by Cantor intersection theorem, both $\cap_n \overline{U_{\alpha|n}}$ and $\cap_n \overline{V_{\alpha|n}}$ are singletons consisting of, say $x(\alpha)$ and $y(\alpha)$, respectively. If $\alpha \neq \beta$, then there exists a n such that $\alpha|n \neq \beta|n$. Since

$$\overline{U_{\alpha|n}} \cap \overline{U_{\beta|n}} = \emptyset = \overline{V_{\alpha|n}} \cap \overline{V_{\beta|n}}$$

$x(\alpha) \neq x(\beta)$ as well as $y(\alpha) \neq y(\beta)$. By (6), $f(x(\alpha)) = y(\alpha)$. Let $C = \{x(\alpha) \in \Sigma : \alpha \in \{0, 1\}^{\mathbb{N}}\}$. Then $|C| = \mathfrak{c}$ and $f|C$ is one to one. Hence $\mathfrak{c} \geq |B| \geq |f(C)| = \mathfrak{c}$. Thus, the proof will be completed.

We now proceed to define systems of open sets $\{U_s : s \in \{0, 1\}^{<\mathbb{N}}\}$ and $\{V_s : s \in \{0, 1\}^{<\mathbb{N}}\}$ in Σ and $\mathbb{R}^n$, respectively, satisfying the above conditions. They are defined by induction on $|s|$. Suppose U_s and V_s have been defined. Since $U_s \cap Y \neq \emptyset$, Y has no isolated points and U_s is open, there exist $a_0 \neq a_1 \in U_s \cap Y$. Since $f|Y$ is one to one and $f(U_s) \subset V_s$, $f(a_0) \neq f(a_1) \in V_s$. Hence, there exist open sets $V_{s\,0}, V_{s\,1}$ containing $f(a_0), f(a_1)$, respectively, of diameters $\leq \frac{1}{2^{|s|+1}}$ such that $\overline{V_{s\,0}}, \overline{V_{s\,1}}$ are disjoint and contained in V_s. Fix a $\epsilon = 0$ or 1. Since $f(a_\epsilon) \in V_{s\,\epsilon}$ and f is continuous at a_ϵ, there exists an open set $U_{s\,\epsilon}$ containing a_ϵ of diameter $\leq \frac{1}{2^{|s|+1}}$ such that $\overline{U_{s\,\epsilon}} \subset U_s$ and $f(\overline{U_{s\,\epsilon}}) \subset V_{s\,\epsilon}$. Since $\overline{V_{s\,0}} \cap \overline{V_{s\,1}} = \emptyset$, $\overline{U_{s\,0}} \cap \overline{U_{s\,1}} = \emptyset$.

This completes the construction and the proof. ∎

Remark 4.5.24 In the above proof we showed that if $f : \Sigma \to \mathbb{R}^n$ is continuous and if $A = f(\Sigma)$ is uncountable, then $|A| = \mathfrak{c}$. Sets $A \subset \mathbb{R}^n$ which are continuous images of Σ are called *analytic*. Subsets of $\mathbb{R}^n$ whose complements in $\mathbb{R}^n$ are analytic are called *coanalytic*. Analytic and coanalytic sets play a fundamental role in the theory of Borel sets.

We now give simple and quite transparent proofs of Hahn decomposition theorem and very useful Radon–Nikodym theorem.

Let Ω be a non-empty set and $\mathcal{A}$ a σ-field on Ω. Then $(\Omega, \mathcal{A})$ is called a *measurable space*. Sets belonging to $\mathcal{A}$ are called *measurable sets*. A *signed measure* on $(\Omega, \mathcal{A})$ is an extended real-valued set function Q on $\mathcal{A}$, i.e., a function $Q : \mathcal{A} \to [-\infty, \infty]$ such that

1. $Q(\emptyset) = 0$.
2. For every sequence $\{A_n\}$ of pairwise disjoint measurable sets,

$$Q(\cup_n A_n) = \sum_n Q(A_n).$$

Let Q be a signed measure on a measurable space $(\Omega, \mathcal{A})$. We list some simple observations that we shall use without specific mention.

(a) Q cannot take both values $+\infty$ and $-\infty$: If possible, suppose there exist measurable sets A and B such that $Q(A) = +\infty$ and $Q(B) = -\infty$. If $A \cap B = \emptyset$, $Q(A \cup B)$ is not defined. If $Q(A \setminus B) = +\infty$, we have disjoint sets $A \setminus B$ and B with measure $+\infty$ and $-\infty$, respectively. This is a contradiction as before. Hence, both $Q(A \setminus B)$ and $Q(B \setminus A)$ are finite. This implies that $Q(A \cap B)$ equals $+\infty$ as well as $-\infty$ which is a contradiction too.

(b) If $Q(A)$ is finite and $B \subset A$, then $Q(B)$ is finite: If possible, suppose $Q(B)$ is infinite. Then

$$Q(A) = Q(B) + Q(A \setminus B).$$

Hence, either $Q(A)$ is infinite or undefined. Set

$$\mathcal{A}|A = \{B \in \mathcal{A} : B \subset A\}$$

and $Q|A$ the restriction of Q to $\mathcal{A}|A$. Then $(A, \mathcal{A}|A, Q|A)$ is a signed-measure space. $\mathcal{A}|A$ is called the *trace σ-field* and $Q|A$ the *trace measure*. Since $Q(A)$ is finite, $Q|A$ is a real-valued signed measure.

(c) Let Q be real valued and $A_n \downarrow A$. Then $\lim_n Q(A_n) = Q(A)$: Since $A_n \setminus A \downarrow \emptyset$, without any loss of generality, we take $A = \emptyset$. Then $A_1 = \cup_{n \geq 1}(A_n \setminus A_{n+1})$. Hence,

$$Q(A_1) = \sum_{n \geq 1}(Q(A_n) - Q(A_{n+1}))$$

is convergent. This implies that $\lim_n Q(A_n) = 0$.

Let $(\Omega, \mathcal{A}, Q)$ be a signed-measure space. An $A \in \mathcal{A}$ is called a *positive set* if $Q(A) > 0$ and for every measurable $B \subset A$, $Q(B) \geq 0$. A positive set with respect to $-Q$ is called a *negative set*. The following result is our main new observation that leads to simpler and much more transparent proofs of Hahn decomposition and Radon–Nikodym theorems.

Proposition 4.5.25 *Let $(\Omega, \mathcal{A}, Q)$ be a signed-measure space. Then every measurable set of positive (negative) measure contains a positive (resp. negative) set.*

Proof Without any loss of generality, we assume that Q does not take value $-\infty$. Take any A with $Q(A) > 0$. If $Q(A) = +\infty$ and every subset of A has measure $+\infty$

4.5 Measure Theory

or 0, A itself is a positive set. If every non-empty, proper subset of A has measure ≤ 0, then $Q(A) \leq 0$ which is a contradiction.

It may happen that every subset of A has measure either $+\infty$ or ≤ 0. Then every subset of A with measure < 0 is a negative set. Now consider

$$\mathcal{N} = \{B \subset A : B \text{ is a negative set}\}$$

and

$$m = \inf\{Q(B) : B \in \mathcal{N}\}.$$

It is easy to see that there is a $C \in \mathcal{N}$ such that $Q(C) = m$. Since Q does not take value $-\infty$, $-\infty < m < 0$. Then every subset of $A \setminus C$ has measure either 0 or $+\infty$ and so a positive subset of A.

Now assume that A has a subset A_0 with $0 < Q(A_0) < \infty$. We set $Q' = Q|A_0$. Then Q' is a real-valued, signed measure on $(A_0, \mathcal{A}|A_0)$.

By transfinite induction, for countable ordinals $\beta < \omega_1$, we define measurable sets A_β, B_β satisfying the following conditions:

(i) $Q'(A_\beta) \geq Q'(A_0) > 0$.
(ii) $B_\beta \subset A_\beta$ and $Q'(B_\beta) < 0$, if such a set exists. Otherwise, we stop because then A_β is a positive subset of A_0.
(iii) $A_{\beta+1} = A_\beta \setminus B_\beta$.
(iv) $A_\delta = \cap_{\beta<\delta} A_\beta$ if δ is a limit ordinal.

We only need to check that sets of the form $A_{\beta+1}$ and A_δ, δ limit, have Q' measure $\geq Q'(A_0) > 0$.

$$Q'(A_{\beta+1}) = Q'(A_\beta) - Q'(B_\beta) > Q'(A_\beta) \geq Q'(A_0) > 0.$$

Since δ is a countable limit ordinal, there is an increasing sequence of countable ordinals $\{\beta_n\}$ with supremum δ. But then $A_{\beta_n} \downarrow A_\delta$. Since Q' is real valued,

$$Q'(A_\delta) = \lim_n Q'(A_{\beta_n}) \geq Q'(A_0) > 0.$$

If possible, suppose the above inductive process does not stop at a countable ordinal. Then, there is a positive integer n such that $I = \{\beta < \omega_1 : Q'(B_\beta) < -\frac{1}{n}\}$ is infinite. Hence, $Q'(\cup_{\beta \in I_1} B_\beta) = -\infty$, where $I_1 \subset I$ is countably infinite. This is a contradiction. Therefore, there is a $\beta < \omega_1$ such that A_β is a positive subset of A.

Next assume that $Q(A) < 0$. Then $Q|A$ is real valued. Now A is a set of positive measure with respect to $-Q|A$. Hence, A contains a positive set with respect to $-Q|A$. This implies that A contains a negative set with respect to Q. ∎

Theorem 4.5.26 (Hahn Decomposition Theorem) *Let $(\Omega, \mathcal{A}, Q)$ be a signed-measure space. Then there exist a positive set Ω^+ and a negative set Ω^- such that $\Omega^+ \cap \Omega^- = \emptyset$ and $\Omega = \Omega^+ \cup \Omega^-$.*

Proof Without any loss of generality, we assume that Q does not take value $-\infty$. Ignoring trivial cases, we also assume that Q takes both positive and negative values. Set
$$\mathcal{N} = \{A \in \mathcal{A} : A \text{ is a negative set}\}$$
and
$$m = \inf\{Q(A) : A \in \mathcal{N}\}.$$

By Proposition 4.5.25, $\mathcal{N} \neq \emptyset$. Take a $\Omega^- \in \mathcal{N}$ such that $Q(\Omega^-) = m$. Set $\Omega^+ = \Omega \setminus \Omega^-$. If Ω^+ contains a measurable set A of negative measure, then by the last Lemma, it contains a negative set B. But then $\Omega^- \cup B$ is a negative set with measure less than m. This contradiction shows that Ω^+ is a positive set. ∎

Theorem 4.5.27 (Radon–Nikodym Theorem) *Let $(\Omega, \mathcal{A})$ be a measurable space and μ, ν probability measures on $(\Omega, \mathcal{A})$ such that $\nu << \mu$. Then there is a random variable Z such that*
$$\forall A \in \mathcal{A}(\nu(A) = \int_A Z d\mu).$$

Proof Set
$$L = \{X \geq 0 : \forall A \in \mathcal{A}(\int_A X d\mu \leq \nu(A))\}.$$

Then following are easily seen:

1. $L \neq \emptyset$.
2. $X, Y \in L \Rightarrow X \vee Y \in L$.
3. $(0 \leq X_n \uparrow X \wedge X_n \in L) \Rightarrow X \in L$.

Now set
$$\alpha = \sup\{\int_\Omega X d\mu : X \in L\}.$$

Using above observations, we easily see that there is a $Z \in L$ such that
$$\int_\Omega Z d\mu = \alpha.$$

We proceed to show that such a Z works. Define
$$\lambda(A) = \nu(A) - \int_A Z d\mu, \ A \in \mathcal{A}.$$

Then λ is a non-negative measure on $(\Omega, \mathcal{A})$. We have to show that $\lambda \equiv 0$. Assume that this is not the case. We shall then arrive at a contradiction.

Observation. $\exists A \in \mathcal{A} \exists k \geq 1(\lambda(A) - \mu(A)/k > 0)$.

4.5 Measure Theory

For otherwise,
$$\forall A \in \mathcal{A} \forall k \geq 1 (\lambda(A) - \mu(A)/k \leq 0).$$

But then $\lambda \equiv 0$ which contradicts our assumption.

Consider the bounded signed measure $Q = \lambda - \mu/k$. By the above observation there exists a measurable set A_0 with $Q(A_0) > 0$. Then by Proposition 4.5.25, A_0 contains a positive set $A \in \mathcal{A}$.

Since $Q(A) > 0$, $\lambda(A) > 0$. Hence, $\nu(A) > 0$. Since $\nu << \mu$, $\mu(A) > 0$. Consider
$$Y = Z + I_A/k \geq 0.$$

Take any $C \in \mathcal{A}$. Then $\int_C Y d\mu$ equals

$$\int_{A^c \cap C} Z d\mu + \int_{A \cap C} Z d\mu + \mu(A \cap C)/k \leq \nu(C).$$

Hence, $Y \in L$.

Clearly,
$$\int_\Omega Y d\mu = \alpha + \mu(A)/k > \alpha.$$

We have arrived at a contradiction by assuming $\lambda \equiv 0$ is false. ■

We now indicate a possibility that well before Lebesgue Cantor might have shown the seed of measure theory. In Section 1 of this chapter, we have presented some observations of Cantor which is of importance to Lebesgue measure. We continue from there now.

What are all the continuous probability measures on $(\mathbb{R}, \mathcal{B}_\mathbb{R})$ singular with respect to the Lebesgue measure λ on $\mathbb{R}$? Remark 4.1.12 allows us to describe a large class of probability measures on $\mathbb{R}$ singular with respect to the Lebesgue measure.

Take any probability P on $(\mathbb{R}, \mathcal{B}_\mathbb{R})$. Define

$$F(x) = P((-\infty, x]), \quad x \in \mathbb{R}.$$

The map $F : \mathbb{R} \to \mathbb{R}$ is called the *distribution function* of the probability measure P. We have

(a) F is non-decreasing.
(b) F is right continuous at all $x \in \mathbb{R}$.
(c) $F(-\infty) = 0$ and $F(\infty) = 1$.

(Check the above three properties.)

A function $F : \mathbb{R} \to \mathbb{R}$ satisfying properties (a) to (c) is called a *distribution function*. Given any distribution function $F : \mathbb{R} \to \mathbb{R}$, for any interval $(a, b]$ we define
$$P((a, b]) = F(b) - F(a).$$

It is well known that P extends to a unique probability measure on the Borel σ-field $\mathcal{B}_{\mathbb{R}}$. It follows that this sets up a one-to-one correspondence between the set of all probability measures on $(\mathbb{R}, \mathcal{B}_{\mathbb{R}})$ and the set of all distribution functions. A probability measure P on $\mathbb{R}$ is called *continuous* if $P(\{x\}) = 0$ for every $x \in \mathbb{R}$. It is easy to check that a probability measure P on $\mathbb{R}$ is continuous if and only if its distribution function is continuous at every $x \in \mathbb{R}$.

Exercise 4.5.28 Show that if $F : \mathbb{R} \to \mathbb{R}$ is a distribution function then its left-hand limit $F(x-)$ exists at every $x \in \mathbb{R}$, $F(x-) \leq F(x)$ for all x and the set of all discontinuity point of F is countable.

Fix a probability measure P on $(\mathbb{R}, \mathcal{B}_{\mathbb{R}})$. Set

$$\mathcal{U} = \{U \subset \mathbb{R} : U \text{ is open and } P(U) = 0\}.$$

Since $\mathbb{R}$ is second countable, there is a countable subcollection $\{U_n\} \subset \mathcal{U}$ such that $\mathcal{U} = \cup_n U_n = V$, say. Then by the countable subadditivity $P(V) = 0$ and it is the largest. The set $C = \mathbb{R} \setminus V$ is the smallest closed subset such that $P(C) = 1$. The set C is called the *topological support* of P.

Exercise 4.5.29 Show that if P is continuous if and only if its topological support is perfect.

To somewhat simplify the rest of the discussion, we take $X = [0, 1]$ and $\mathcal{B} = \mathcal{B}_{[0,1]}$, the Borel σ-field of the closed unit interval.

From now onward, we fix a perfect set $C \subset [0, 1]$ containing 0 and 1. In addition assume that

$$[0, 1] \setminus C = \cup_n (a_n, b_n)$$

such that for $n \neq m$, $(a_n, b_n) \cap (a_m, b_m) = \emptyset$ and $\sum_n (b_n - a_n) = 1$. Hence, the Lebesgue measure of C equals 0. In particular, C contains no non-empty open interval.

Let P be a continuous probability measure on $(X, \mathcal{B})$ with topological support C. Then P is singular with respect to the Lebesgue measure. Its distribution function F has the following additional properties:

 (i) F is constant on each non-empty open interval disjoint from C.
 (ii) $F|C$ is strictly increasing.
 (iii) $F(0) = 0$ and $F(1) = 1$.

Conversely, if F is a distribution function satisfying (i) to (iii) and P corresponding probability measure with distribution function F, then P is singular with respect to the Lebesgue measure. Thus, Remark 4.1.12 allows us to describe all continuous probabilities on $(X, \mathcal{B})$ with topological support C which are singular with respect to the Lebesgue measure.

Remark 4.5.30 I am very grateful to M. G. Nadkarni for drawing my attention to [6] and to his very interesting article [7]. All the materials on the general construction of continuous probability measures on $\mathbb{R}$ singular with respect to the Lebesgue measure as discovered by Cantor are from [7]. In this beautiful note, Nadkarni argues quite convincingly that Cantor had sown the seed of measure and integral which finally culminated into Lebesgue's invaluable development of Lebesgue measure and Lebesgue integration.

References

1. S.M. Srivastava, *A Course on Borel Sets, GTM 180* (Springer, New York, 1997)
2. P. Erdös, An interpolation problem associated with the continuum hypothesis. Michigan Math. J. **11**, 9–10 (1964)
3. J.L. Kelly, The Tychonoff product theorem implies the axiom of choice. Fund. Math. **37**, 75–76 (1950)
4. A.H. Stone, Paracompactness and product space. Bull. Amer. Math. Soc. **54**, 977–982 (1948)
5. I.K. Rana, *An Introduction to Measure and Integration*, 2nd edn. Graduate Studies in Mathematics, vol. 45 (American Mathematical Society, 2002)
6. G. Cantor, Gesammelte Abhandlungen, in *Varlagbuchhandlungen Abhandlungen*, ed. by E. Zermelo (Georg Olms, Reprint 1962, Original 1932)
7. M.G. Nadkarni, Did Cantor sow the seed of measure and integral? Ramanujan Mathe. Soc. **15**(4), 93–99 (2006). (March)

Chapter 5
Banach–Tarski Paradox

In Exercise 4.5.16, we showed that there is no translation invariant, countably additive measure μ on $\mathcal{P}(\mathbb{R})$ with $\mu([0, 1]) = 1$. This showed that we cannot extend the notion of "length" to all subsets of $\mathbb{R}$ provided we accept that "length" should be countably additive, i.e., the length of a countable disjoint union of sets should be the sum of the lengths of the sets from the countable family.

The restriction of countable additivity of "length" may appear to be a strict requirement. Question arises—what happens if we relax the requirement of countable additivity to finite additivity. *Banach (4.5.17) showed that if $n = 1$ or 2, then there is a finitely additive measure μ on $\mathcal{P}(\mathbb{R}^n)$ such that $\mu([0, 1]^n) = 1$ and whenever $A, B \subset \mathbb{R}^n$ are congruent, $\mu(A) = \mu(B)$.* The proof uses functional analysis and is beyond the scope of this note.

In a very painstaking work, Hausdorff showed that Banach's result is not true for $n > 2$. In Sect. 5.1 of this chapter, we shall present a complete proof of Hausdorff's theorem for $n = 3$. Since not much of set theory is required to generalize Hausdorff's theorem for $n > 3$, we shall omit that part.

Using Hausdorff's proof Banach and Tarski [1] showed that the closed unit ball D_3 in $\mathbb{R}^3$ can be partitioned into finitely many sets that can be so rearranged that one gets two balls each equal in size to D_3. This is known as Banach–Tarski paradox. We present Banach–Tarski paradox in Sect. 5.2.

Hausdorff uses AC in his proof. Probably so far in this note the reader may not have found any serious counterintuitive consequence of AC. This consequence of AC is definitely counterintuitive and advocates some caution in using AC. In Sect. 5.1, we shall give a proof of Hausdorff's theorem. Banach–Tarski paradox is presented in Sect. 5.2.

5.1 Hausdorff's Theorem

Theorem 5.1.1 (Hausdorff) *There is no finitely additive measure μ on $\mathcal{P}(\mathbb{R}^3)$ such that $\mu([0.1]^3) = 1$ and whenever $A, B \subset \mathbb{R}^3$ are congruent, $\mu(A) = \mu(B)$.*

Proof We assume that such a μ exists. We shall arrive at a contradiction. Set

$$D_3 = \{x \in \mathbb{R}^3 : |x| \leq 1\},$$

the closed unit ball of $\mathbb{R}^3$. We make a few observations first which will be used subsequently without any mention.

(1) Let a be some element in $\mathbb{R}^3$. By invariance under translation, for every $b \in \mathbb{R}^3$, $\mu(\{a\}) = \mu(\{b\})$. If possible suppose $\mu(\{a\}) > 0$. Since $[0, 1]^3$ contains a countably infinite set, we see that for every positive integer k

$$1 = \mu([0, 1]^3) \geq k\mu(a).$$

This is a contradiction. It shows that for every $a \in \mathbb{R}^3$, $\mu(\{a\}) = 0$.

(2) By invariance under congruence $\mu(C) = 1$ for every closed cube in $\mathbb{R}^3$ with all sides of unit length.

(3) Let $A \subset \mathbb{R}^3$ be bounded. Then there is a $m \geq 1$ and closed cubes $C_1, \ldots, C_m$ with all sides of unit length such that $A \subset \cup_{i=1}^m C_i$. Since μ is finitely additive, it is finitely subadditive. Therefore, $\mu(A) \leq m < \infty$. In particular, $\mu(D_3)$ is finite.

(4) Let $z \in S^2$, the unit sphere in $\mathbb{R}^3$, and $(0, z]$ the line segment joining 0 and z but not including 0. Then $\mu((0, z]) = 0$. Suppose not. Then by invariance under rotation, for every $w \in S^2$, $\mu((0, w]) = \mu((0, z]) > 0$. However, S^2 being uncountable, D_3 contains infinitely many such line segments. Further they are pairwise disjoint. Hence, for every $n \geq 1$,

$$\infty > \mu(D_3) \geq n\mu((0, z]).$$

This is a contradiction. It follows that for every $z \in S^2$, $\mu((0, z]) = 0$.

(5) Let $Q \subset S^2$ be countable. Then there exists a sequence $\{Q_n\}$ of pairwise disjoint countable subsets of S^2 each a rotation of Q. Arguing as before we get a contradiction if

$$\mu(\cup_{z \in Q}(0, z]) > 0.$$

Hence,

$$\mu(\cup_{z \in Q}(0, z]) = 0.$$

Exercise 5.1.2 Show that $\mu(D_3) > 0$.

Hausdorff's proof is based on his following main lemma.

Lemma 5.1.3 *The sphere S^2 is the union of four pairwise disjoint subsets Q, R, S, and T such that Q is countable and each of R, S, T, and $S \cup T$ is the image of others under a rotation around a line passing through origin.*

5.1 Hausdorff's Theorem

We assume the main lemma and complete the proof of Hausdorff's theorem first. Define

$$Q_0 = \cup_{z \in Q}(0, z],$$

$$R_0 = \cup_{z \in R}(0, z],$$

$$S_0 = \cup_{z \in S}(0, z],$$

and

$$T_0 = \cup_{z \in T}(0, z].$$

Then Q_0, R_0, S_0, and T_0 are pairwise disjoint with union $D_3 \setminus \{0\}$. Further, each of R_0, S_0, T_0, and $S_0 \cup T_0$ is the image of others under a rotation around a line passing through origin. As observed above $\mu(Q_0) = 0$. By the invariance of μ under rotations,

$$\mu(R_0) = \mu(S_0) = \mu(T_0) = \mu(S_0 \cup T_0) < \infty.$$

By this together with the finite additivity of μ, we have

$$0 < \mu(D_3 \setminus \{0\}) = \mu(Q_0) + \mu(R_0) + \mu(S_0) + \mu(T_0) = 3\mu(R_0).$$

Thus, $\mu(R_0) > 0$. However, since $S_0 \cup T_0$ is the image of R_0 under a rotation, we have

$$\mu(R_0) = \mu(S_0 \cup T_0) = \mu(S_0) + \mu(T_0) = 2\mu(R_0)$$

implying that $\mu(R_0) = 0$. We have arrived at a contradiction.

Thus, Hausdorff's theorem will be proved if we prove his main lemma.

Let $u \in S^2$ be a point in the second quadrant of $X - Z$ plane of $\mathbb{R}^3$ such that the angle between the line segments $[0, e_3]$ and $[0, u]$ is θ, $0 < \theta < \pi/2$, where $e_3 = (0, 0, 1)$. Let φ be the anticlockwise rotation around positive Z-axis by π and ψ the anticlockwise rotation around the line segment $\overrightarrow{0\,u}$ in the direction of u by $120°$. Then $\varphi^2 = 1$ and $\psi^3 = 1$, where 1 denotes the identity rotation.

Let G be the subgroup of the group of congruences under the composition map generated by φ and ψ. Note that each transformation in G is a rotation around a line passing through the origin. Hence, G acts on S^2.

Warning. *Under the group notation for ρ, $\sigma \in G$, $\rho \cdot \sigma$ stands for the composition $\sigma \circ \rho$.*

Any non-identity element σ can be expressed as

$$\sigma = \varphi^{i_1} \cdot \psi^{j_1} \cdot \varphi \cdots \varphi \psi^{j_n} \cdot \varphi^{i_{n+1}}, \tag{*}$$

where $i_1, i_{n+1} = 0$ or 1 and $j_1, \ldots, j_n = 1$ or 2. In this case, we say that this representation of σ is in the *reduced form*.

While multiplying any two rotations, we can delete two consecutive appearances of φ and any two consecutive appearances of ψ and ψ^2 to express the product of two rotations in G in reduced form as in (∗).

It may happen that for some choice of θ, a $\sigma \in G$ with representation as in (∗) of positive length may become the identity rotation. We show that we can choose $0 < \theta < \pi/2$ in such a way that this is not possible.

We first show that there is a θ, $0 < \theta < \pi/2$, such that no rotation α of the form

$$\alpha = \varphi \cdot \psi^{j_1} \cdots \varphi \cdot \psi^{j_n},$$

where $j_1, \ldots, j_n = 1$ or 2, equals 1. Fix $(x, y, z) \in \mathbb{R}^3$. One can calculate and see the following.

If $\varphi \cdot \psi(x, y, z) = (x_1, y_1, z_1)$, then

$$x_1 = \frac{1}{2}x \cos \theta + \frac{\sqrt{3}}{2}y - \frac{1}{2}z \cos \theta,$$

$$y_1 = -\frac{\sqrt{3}}{2}x \cos \theta + \frac{1}{2}y + \frac{\sqrt{3}}{2}z \cos \theta$$

and

$$z_1 = x \sin \theta + z \cos \theta.$$

($\varphi \cdot \psi$ is the composition of three functions $h \circ g \circ f$ where

(1) f is the rotation around Z-axis by angle π. Therefore, $g(x, y, z) = (-x, -y, z)$.

(2) g is the anticlockwise rotation of $X - Z$ plane by angle θ. This is the same as the anticlockwise rotation around Y-axis by angle θ. Hence,

$$f(r \cos(\varphi), y, r \sin(\varphi)) = (r \cos(\varphi + \theta), y, r \sin(\varphi + \theta)).$$

(3) h is the anticlockwise rotation around Z-axis by angle $2\pi/3$. Then

$$h(r \cos \phi, r \sin \phi, z) = (r \cos(2\pi/3 + \varphi), r \sin(2\pi/3 + \varphi), z).)$$

Similarly, one sees that if $\varphi \cdot \psi^2(x, y, z) = (x_2, y_2, z_2)$, then

$$x_2 = \frac{1}{2}x \cos \theta - \frac{\sqrt{3}}{2}y - \frac{1}{2}z \cos \theta,$$

$$y_2 = \frac{\sqrt{3}}{2}x \cos \theta + \frac{1}{2}y - \frac{\sqrt{3}}{2}z \cos \theta$$

and

$$z_2 = x \sin \theta + z \cos \theta.$$

5.1 Hausdorff's Theorem

(Note that $\varphi \cdot \psi^2 = h' \circ g \circ f$, where h' is the clockwise rotation around Z-axis by angle $2\pi/3$. We then have

$$h'(r\cos\phi, r\sin\phi, z) = (r\cos(\varphi - 2\pi/3), r\sin(\varphi - 2\pi/3), z).)$$

Suppose $\alpha(0, 0, 1) = (0, 0, 1)$. Then there exist real numbers $a_0, \ldots, a_n$ with $a_n \neq 0$ with

$$\sum_{i=0}^{n} a_i \cos^i \theta = 1,$$

i.e., $\cos\theta$ is a root of a polynomial of the form

$$\sum_{i=0}^{n} a_i X^i - 1.$$

But as n varies from 1 to ∞, there are only countably many possible roots of a polynomial of the above form. Hence, there exists a θ, $0 < \theta < \pi/2$, such that $\cos\theta$ is not equal to any of these countably many roots.

We choose and fix such a θ. Hence, no rotation other than 1 in G of the form α equals 1.

We show that for no $\sigma \in G$ in the reduced form other than 1, $\sigma = 1$. Clearly, none of φ, ψ and ψ^2 equals 1.

Suppose a rotation of the form

$$\beta = \psi^{j_1} \cdot \varphi \cdots \psi^{j_n} \cdot \varphi = 1.$$

Then

$$1 = \varphi \cdot \beta \cdot \varphi = \varphi \cdot \psi^{j_1} \cdot \varphi \cdots \psi^{j_n}.$$

However, we have seen that such a rotation $\alpha \neq 1$.

Next suppose that a rotation of the form

$$\gamma = \varphi \cdot \psi^{j_1} \cdots \varphi \cdot \psi^{j_n} \cdot \varphi = 1.$$

Then

$$1 = \varphi \cdot \gamma \cdot \varphi = \psi^{j_1} \cdot \varphi \cdots \varphi \cdot \psi^{j_n} = \delta,$$

say.

Using the fact that no rotation of the form $\alpha = 1$, we now show that no rotation of the form $\delta = 1$. Suppose there is a rotation of the form $\delta = 1$. Choose one such δ of minimum length. Say

$$\delta = \psi^{j_1} \cdot \varphi \cdots \varphi \cdot \psi^{j_n}.$$

Then $n > 1$ and the length of $\delta \geq 3$.

Case: 1. $j_1 + j_n \neq 3$. Then

$$1 = \psi^{-j_1} \cdot (\psi^{j_1} \cdot \varphi \cdots \varphi \cdot \psi^{j_n+j_1}).$$

Since $j_1 + j_n \neq 3$, the rotation on the right is of the form α. This gives us a contradiction.

Case: 2. $j_1 + j_n = 3$. Then

$$1 = \varphi \cdot \psi^{j_n} \cdot (\psi^{j_1} \cdot \varphi \cdots \varphi \cdot \psi^{j_n}) \cdot \psi^{j_1} \cdot \varphi$$

is a rotation of the form δ of smaller length. This contradicts the minimality of the length of δ.

Now we have proved that our choice of θ is such that no $\rho \in G$ other than 1 in reduced form is identity.

Since the group G is countable and each rotation in $G \setminus \{1\}$ fixes exactly two points of S^2, the set $Q \subset S^2$ defined by

$$Q = \{z \in S^2 : \exists 1 \neq \rho \in G(\rho(z) = z)\}$$

is countable.

Then G acts on $S^2 \setminus Q$: Suppose not. Then there is a $z_0 \in S^2 \setminus Q$ and $1 \neq \sigma \in G$ such that $\sigma(z_0) \in Q$. Get $1 \neq \rho \in G$ such that $\rho(\sigma(z_0)) = \sigma(z_0)$. It follows that $\sigma^{-1}(\rho(\sigma(z_0))) = z_0$. Since $z_0 \notin Q$, we must have $\sigma \cdot \rho \cdot \sigma^{-1} = 1$. Hence, $\sigma \cdot \rho = \sigma$ implying $\rho = 1$. This is a contradiction.

For each $x \in S^2 \setminus Q$, let $O_G(x)$ denote the orbit of x under the action of G. By AC, there is a set $X \subset S^2 \setminus Q$ such that $|O_G(x) \cap X| = 1$ for all $x \in S^2 \setminus Q$.

Main Claim. *There is a partition of G into three non-empty sets A, B, and C satisfying the following two conditions:*

(I) $\rho \in A \Leftrightarrow \rho \cdot \varphi = \varphi \circ \rho \in B \cup C$.
(II) $\rho \in A \Leftrightarrow \rho \cdot \psi = \psi \circ \rho \in B \Leftrightarrow \rho \cdot \psi^2 = \psi^2 \circ \rho \in C$.

Assuming the main claim, we complete the proof first. We define

$$R = \cup_{\rho \in A} \rho(X),$$

$$S = \cup_{\rho \in B} \rho(X),$$

and

$$T = \cup_{\rho \in C} \rho(X).$$

Note that $S^2 \setminus Q = R \cup S \cup T$.

Claim: R, S, and T are pairwise disjoint.

5.1 Hausdorff's Theorem

$\underline{R \cap S = \emptyset}$: Suppose not. Then there exist $\rho_1 \in A$, $\rho_2 \in B$, $a \in X$, and $b \in X$ such that $\rho_1(a) = \rho_2(b)$. Then $\rho_2^{-1} \cdot \rho_1(a) = b$. Thus, a and b belong to the same orbit. Since $a, b \in X$, $a = b$. On the other hand, since $A \cap B = \emptyset$, $\rho_1 \neq \rho_2$. Hence, $\rho_2^{-1} \circ \rho_1 \neq 1$. Whenever $\rho_2^{-1} \circ \rho_1 \neq 1$, its two fixed points are in Q. Since $\rho_2^{-1} \circ \rho_1(a) = b$, it follows that $a \neq b$. We have arrived at a contradiction.

Similarly we prove that $R \cap T = \emptyset = S \cap T$.

$\varphi(R) = S \cup T$: Let $a \in R$. Get $\rho \in A$ and $b \in X$ such that $\rho(b) = a$. Then $\varphi(a) = \varphi(\rho(b)) = \rho \cdot \varphi(b)$. By (I). $\rho \cdot \varphi \in B \cup C$. Hence, $\varphi(R) \subset S \cup T$.

Conversely, let $a \in S$. Then there exist $\rho \in B$ and $b \in X$ such that $a = \rho(b)$. Take $\sigma = \rho \cdot \varphi^{-1}$. By (I), $\sigma \in A$. Therefore, $R \ni \sigma(b)$ and $\varphi(\sigma(b)) = \rho(b) = a$. Thus, $S \subset \varphi(R)$. Similarly, it follows that $T \subset \varphi(R)$. It follows that $S \cup T = \varphi(R)$.

Arguing similarly and using (II) only, we show that $\psi(R) = S$ and $\psi^2(R) = T$.

It remains to prove our main claim, i.e., we are required to partition G into three sets A, B and C such that for any word

$$\sigma = \varphi^{i_0} \cdot \psi^{j_0} \cdot \varphi \cdots \varphi \psi^{j_{n-1}} \cdot \varphi^{i_n}, \qquad (**)$$

where $i_0, i_n = 0$ or 1 and $j_0, \ldots, j_{n-1} = 1$ or 2, in G the following two conditions are satisfied.

(I) $\sigma \in A \Leftrightarrow \sigma \cdot \varphi \in B \cup C$.
(II) $\sigma \in A \Leftrightarrow \sigma \cdot \psi \in B \Leftrightarrow \sigma \cdot \psi^2 \in C$.

For each $n \geq 1$, set

$$G_n = \{\sigma : \sigma \text{ satisfies } (**) \text{ and the length of } \sigma \leq n\}.$$

For instance, $G_1 = \{1, \varphi, \psi, \psi^2\}$. By induction on $n \geq 1$, we shall define a partition A_n, B_n, C_n of G_n such that for $n = 1, 2, 3, \ldots$ the following conditions are satisfied.

1. $A_{n+1} \cap G_n = A_n$, $B_{n+1} \cap G_n = B_n$ and $C_{n+1} \cap G_n = C_n$.
2. Conditions (I) and (II) are satisfied for words in G_n.

Take $A_1 = \{1\}$, $B_1 = \{\varphi, \psi\}$, and $C_1 = \{\psi^2\}$. If $\sigma = 1 \in A_1$, then clearly $1 \cdot \varphi$, $1 \cdot \psi \in B_1$ and $1 \cdot \psi^2 \in C_1$. Conversely, let $\sigma \in G_1$ and $\sigma \cdot \varphi \in B_1 \cup C_1$. If $\sigma \cdot \varphi \in B_1 \cup C_1$, we must have $\sigma \cdot \varphi = \varphi$. Then $\sigma = 1 \in A_1$. Similarly, $\sigma \cdot \psi \in B_1$ implies $\sigma = 1 \in A_1$. It is clear that $\sigma \cdot \psi \in C_1 \Leftrightarrow \sigma = \psi \in B_1$. We have proved that for words in G_1, A_1, B_1, and C_1 satisfy (I) and (II).

Now assume that A_n, B_n, C_n is a partition of G_n such that for words in G_n, A_n, B_n, and C_n satisfy (I) and (II). Take any $\rho \in G_{n+1} \setminus G_n$ in reduced form. Then ρ can be one of the following three types:

(1) $\rho = \sigma \cdot \varphi$,
(2) $\rho = \sigma \cdot \psi$,
(3) $\rho = \sigma \cdot \psi^2$,

where σ is of length n and in reduced form. There are three possibilities for σ:

(a) $\sigma = \cdots \varphi$,
(b) $\sigma = \cdots \psi$,
(c) $\sigma = \cdots \psi^2$.

Since ρ is in reduced form and of length $n+1$, none of the pairs (1) and (a) or (2) and (c) or (3) and (b) or (3) and (c) can hold simultaneously.

Suppose (1) is the case, i.e., $\rho = \sigma \cdot \varphi$. Then put ρ in $B_{n+1} \cup C_{n+1}$ if $\sigma \in A_n$. If $\sigma \notin A_n$, then we put $\rho \in A_{n+1}$.

Now assume that (2) holds, i.e., $\rho = \sigma \circ \psi$. If (a) holds and $\sigma \in A_n$, then put $\rho \in B_{n+1}$. If (a) holds $\sigma \in B_n$, then put $\rho \in C_{n+1}$. If (a) holds and $\sigma \in C_n$, then put $\rho \in A_{n+1}$.

Finally, let $\rho = \sigma \circ \psi^2$. Then (a) holds and we put $\rho \in C_{n+1}$.

Define
$$A = \cup_n A_n, \ B = \cup_n B_n \text{ and } C = \cup_n C_n.$$

Then A, B, C is a partition of the group G satisfying (I) and (II). This completes the proof of the main lemma of Hausdorff. ∎

5.2 Banach–Tarski Paradox

Let $X, Y \subset \mathbb{R}^3$. We say that X and Y are equal in size, and write $X \approx Y$, if both X and Y are disjoint union of same number of sets $X_1, \ldots, X_n$ and $Y_1, \ldots, Y_n$, respectively, such that for each $1 \leq i \leq n$, X_i and Y_i are congruent. The following is an easy exercise.

Exercise 5.2.1 Show the following.

1. Let $X_1 \cap X_2 = \emptyset = Y_1 \cap Y_2$ and $X_i \approx Y_i$, $i = 1, 2$. Then $X_1 \cup X_2 \approx Y_1 \cup Y_2$.
2. $\approx$ is an equivalence relation on $\mathcal{P}(\mathbb{R}^3)$.

Lemma 5.2.2 *Let $Z \subset Y \subset X \subset \mathbb{R}^3$ and $Z \approx X$. Then $Y \approx X$.*

Proof Let X and Z be disjoint unions of $X_1, \ldots, X_n$ and $Z_1, \ldots, Z_n$, respectively, such that for each $1 \leq i \leq n$, there is a congruence $f_i : X_i \to Z_i$. Set $f = \cup_{i=1}^n f_i : X \to Z$. For each $n \in \mathbb{N}$, define
$$A_n = f^n(X) \ \& \ B_n = f^n(Y).$$

Then $A_0 \supset A_1 \supset A_2, \ldots, B_0 \supset B_1 \supset B_2, \ldots$ and for every n, $B_n \subset A_n$. Define $A = \cup_n (A_n \setminus B_n)$. Then $A \approx f(A)$. Further, $f(A) \cap (X \setminus A) = \emptyset$. Note that $X = A \cup (X \setminus A)$ and $Y = f(A) \cup (X \setminus A)$. Hence, by the above exercise, $Y \approx X$. ∎

5.2 Banach–Tarski Paradox

Theorem 5.2.3 (Banach–Tarski Paradox) *The closed unit ball D_3 of $\mathbb{R}^3$ is the disjoint union of two sets X and Y such that $X \approx D_3 \approx Y$.*

Proof We shall be following the same notation as in the proof of Hausdorff's Theorem 5.1.1. We showed that

$$D_3 = \{\overline{0}\} \cup Q_0 \cup R_0 \cup S_0 \cup T_0,$$

where $\overline{0}$ denotes the origin, all the above five sets are pairwise disjoint, $Q = Q_0 \cap S^2$ is countable and R_0, S_0, T_0, and $S_0 \cup T_0$ are pairwise congruent. Then

$$R_0 \approx S_0 \approx T_0 \approx R_0 \cup S_0 \cup T_0 :$$

We have $S_0 \approx R_0$ and $T_0 \approx S_0 \cup T_0$. Hence, by Exercise 5.2.1 (1), $S_0 \cup T_0 \approx R_0 \cup S_0 \cup T_0$. We further have $R_0 \approx S_0 \cup T_0$. Since $\approx$ is an equivalence relation, we get $R_0 \approx R_0 \cup S_0 \cup T_0$. Similarly, we show that $S_0, T_0 \approx R_0 \cup S_0 \cup T_0$.
Set $X = R_0 \cup Q_0 \cup \{\overline{0}\}$ and $Y = D_3 \setminus X = S_0 \cup T_0$.
Since $R_0 \approx R_0 \cup S_0 \cup T_0$, by Lemma 5.2.2,

$$X \approx R_0 \cup S_0 \cup T_0 \cup Q_0 \cup \{\overline{0}\} = D_3.$$

Since Q and the group G generated by φ and ψ considered in the proof of Hausdorff's theorem are countable, there is a rotation σ around a line passing through the origin and not in G such that $Q \cap \sigma(Q) = \emptyset$, i.e., $\sigma(Q) \subset R \cup S \cup T$.
Since $T_0 \approx R_0 \cup S_0 \cup T_0$, we see that there is a countable $Q' \subset T_0 \cap S^2$ such that $Q'_0 \subset T_0$, where

$$Q'_0 = \cup_{z \in Q'} (0, z].$$

Choose any point $p \in T_0 \setminus Q'_0$. Clearly,

$$X = R_0 \cup Q_0 \cup \{\overline{0}\} \approx S_0 \cup Q'_0 \cup \{p\} \approx D_3.$$

Since

$$S_0 \cup Q'_0 \cup \{p\} \subset S_0 \cup T_0 = Y \subset D_3,$$

by Lemma 5.2.2 it follows that $Y \approx D_3$. ∎

Remark 5.2.4 In the proof of the above theorem, we have given a paradoxical decomposition of the unit ball using five pieces. However, it can be shown that the paradoxical decomposition cannot be achieved using four pieces. Moreover, as opposed to the use of AC in the proof of the Banach–Tarski paradox, a proof of the paradox using Hahn–Banach theorem, which is known to be weaker than AC, yields a paradoxical decomposition using six pieces. In the same spirit employing Hahn–Banach theorem yields the existence of a non-measurable set while avoiding the full strength of AC [2].

Remark 5.2.5 In spite of such counterintuitive paradoxes, such as the Banach–Tarski paradox, popping up in ZFC, it turns out that life without AC is by no means a bed of roses in terms of the absence of such paradoxes popping up in ZF—without AC. In fact, an attempt at eliminating the Banach–Tarski paradox (respectively AC) yields the following paradox which is even more startling than the Banach–Tarski paradox—and without AC.

The reals can be divided into non-empty classes so that there are strictly more classes than there are reals!! [3].

I thank the unanimous referee for drawing my attention to the last two remarks.

References

1. S. Banach, A. Tarski, Sur la décomposition des ensembles de points en parties respectivement congruentes. Fund. Math. **6**, 244–277 (1924)
2. G. Tomkowicz, S. Wagon, *The Banach-Tarski Paradox*. Encyclopedia of Mathematics and its Applications (Cambridge University Press, New York, 2016)
3. A.D. Taylor, S. Wagon, A paradox arising from the elimination of a paradox. Amer. Math. Monthly **126**(4), 306–318 (2019)

Chapter 6
Preliminary Concepts and Terminologies from Logic

In this chapter, we give some preliminary concepts and terminologies from first-order logic. Readers are advised to see [1] for a more detailed study.

First-order languages for all theories have common logical symbols. They are

(a) **variables:** An infinite sequence of distinct symbols $x_0, x_1, x_2, \ldots$.
(b) **logical connectives:** $\neg$ to be read as *negation* and $\vee$ to be read as *or*.
(c) **logical quantifier:** $\exists$ to be read as *there exists*.
(d) **equality symbol:** $=$, a binary relation symbol.

Formally variables are fixed symbols with a fixed enumeration, but we shall use any English letter (both capital and small) with or without suffixes for variables. This is to maintain some informality in our presentation. The above symbols are called *logical symbols*.

Depending on the theory, say theory of groups, theory of ordered fields, number theory, set theory, etc., under consideration, a first-order language also consists of the following collection of symbols:

(d) **relation or predicate symbols:** For each $n \geq 1$, a set of n-ary relation symbols. Notice that we do not insist upon these sets to be non-empty. The language of a theory need not have any relation symbol.
(e) **function symbols:** For each $n \geq 1$, a set of n-ary function symbols.
(f) **constant symbols:** a set of constant symbols.

These symbols are called *non-logical symbols*. Though parentheses (,), {, }, [, and] are not a part of any first-order language. We use them to avoid confusions and for unique readability.

(1) The language of group theory has a constant symbol e for group identity and a binary (2-ary) function symbol $\cdot$ for the group operation. Sometimes we shall also use the symbol $+$ instead of $\cdot$. Generally this is done when the group under consideration is abelian.

(2) The language for the theory of ordered fields has two constant symbols 0 and 1 for additive and multiplicative identities, respectively, two binary function symbols $+$ and $\cdot$ for field addition and multiplication, respectively, and one binary relation symbol $<$.

(3) The language for set theory has only one non-logical symbol $\in$, a binary relation symbol, to be read as *belongs to*.

A finite sequence of logical and non-logical symbols is called an *expression*.

We fix a first-order language L. The set of all *terms* τ is the smallest set of expressions which contains each variable and each constant symbol and whenever $t_1, \ldots, t_n \in \tau$ and f a n-ary function symbol, $f(t_1, \ldots, t_n) \in \tau$. We shall write $t[x_0, \ldots, x_{n-1}]$ to indicate that t is a term in which no variable other than $x_0, \ldots, x_{n-1}$ occur. In a term no variable may occur. Such terms are called *variable-free terms*.

The set of all *atomic formulas* are expressions of the form $t_1 = t_2$, where t_1 and t_2 terms and $p(t_1, \ldots, t_n)$ where p is a n-ary relation symbol and $t_1, \ldots, t_n$ are terms.

The set of all *formulas* is the smallest set $\mathcal{F}$ of expressions containing all atomic formulas and whenever $\varphi, \psi \in \mathcal{F}$ and v is a variable, $\neg\varphi, \varphi \vee \psi$, and $\exists v \varphi$ are in $\mathcal{F}$. If φ is a formula, the set S_φ of all its *subformulas* is the smallest set of formulas that contains φ and whenever a formula of the form $\neg\psi$ or $\psi \vee \eta$ or $\exists v \psi$ is in S_φ, so is ψ and η.

An occurrence of a variable v in a formula φ is called a *bound occurrence* if it occurs in a subformula of the form $\exists v \psi$. Otherwise, the occurrence of v is called a *free occurrence*. Let t be a term, x a variable, and φ a formula. We say that t is *Substitutable* for x in t if for each variable v occurring in t, no subformula of φ of the form $\exists v \psi$ contains an occurrence of x that is free in φ. We write $\varphi[x_0, \ldots, x_{n-1}]$ to indicate that no variable other than $x_0, \ldots, x_{n-1}$ has a free occurrence in φ. If terms $t_0, \ldots, t_{n-1}$ are substitutable for $x_0, \ldots, x_{n-1}$, respectively, in $\varphi[x_0, \ldots, x_{n-1}]$, then

$$\varphi_{x_0,\ldots,x_{n-1}}[t_0, \ldots, t_{n-1}]$$

will denote the formula obtained by simultaneously replacing each free occurrence of $x_0, \ldots, x_{n-1}$ in φ by $t_0, \ldots, t_{n-1}$, respectively. A formula in which $\exists$ does not occur is called an *open formula* or a *quantifier-free formula*. A formula φ is called a *closed formula* or a *sentence* if no variable has a free occurrence in φ.

We define

$\varphi \wedge \psi$ to be the formula $\neg(\neg\varphi \vee \neg\psi)$,

$\varphi \to \psi$ to be the formula $\neg\varphi \vee \psi$,

$\varphi \leftrightarrow \psi$ to be the formula $(\varphi \to \psi) \wedge (\psi \to \varphi)$,

and

$\forall v \varphi$ to be the formula $\neg \exists v \neg \varphi$.

A *first-order theory* or simply a *theory* T consists of a first-order language $L(T)$ or L (when there is no scope for confusion) and a set of sentences of L, whose elements are called *non-logical axioms* of T.

A *structure for a first-order language* L consists of a non-empty set M, and for each constant symbol c of L an element $c^M \in M$, for each n-ary relation symbol p of L a n-ary relation $p^M \subset M^n$ on M and for each k-ary function symbol f a k-ary

6 Preliminary Concepts and Terminologies from Logic

function $f^M : M^k \to M$ from M^k to M. c^M, p^M, and f^M are called *interpretations* of non-logical symbols c, p, and f, respectively.

For each variable-free term t we now proceed to define the *value* t^M of t in M. This we shall define by induction on the length of t. If the length of t is one, then t is a constant symbol, say c, of L. In this case c^M is already given by the definition of the structure M. Now suppose $t = f(t_1, \ldots, t_n)$, f a n-ary function symbol and t a variable-free term of L. Then $t_1, \ldots, t_n$ are variable-free terms. Suppose $t_1^M, \ldots, t_n^M$ have been defined already. Then we define $t^M = f^M(t_1^M, \ldots, t_n^M)$.

Let φ be a sentence in the language L. We now proceed to define when φ is true or satisfied in M. Let L^M be the language obtained from L by adding each $a \in M$ as a new constant symbol of L^M. M can be canonically thought of as a structure of L^M by interpreting each new constant symbol $a \in M$ by a itself.

Let φ be a sentence in the language L^M. We now proceed to define when φ is true or satisfied in M. This we shall do by induction on the length of φ. If φ is not true in M, it is false in M. Let φ be the atomic formula $t_1 = t_2$ where t_1 and t_2 are terms. Since $t_1 = t_2$ is a sentence, we say that $t_1 = t_2$ is true in M if $t_1^M = t_2^M$. Otherwise, $t_1 = t_2$ is false in M. If φ is an atomic sentence $p(t_1, \ldots, t_n)$, then $t_1, \ldots, t_n$ are variable-free terms. In this case, we say that φ is true in M if the n-tuple $(t_1^M, \ldots, t_n^M)$ satisfies the relation p^M, i.e., $(t_1^M, \ldots, t_n^M) \in p^M$. In this case in logic it is customary to write $p^M(t_1^M, \ldots, t_n^M)$.

Now let ψ be a sentence in L^M and φ be $\neg \psi$. Then φ is true in M if and only if ψ is false in M. Let ψ and η be sentences of L^M and φ be $\psi \vee \eta$. Then φ is true in M if and only if either ψ or η is true in M. Finally, let φ be a sentence of L^M of the form $\exists x \psi$. Since φ is closed, at most one variable x can be free in ψ. Then φ is true in M if and only if there is an $a \in M$ such that $\psi_x(a)$ is true in M.

Let $\varphi[x_0, \ldots, x_{n-1}]$ be such that x_{n-1} has a free occurrence in φ. We call the sentence $\forall x_0 \cdots \forall x_{n-1} \varphi$ the *closure* of φ. We say that φ is true in M if and only if its closure is true in M. Note that $\varphi[x_0, \ldots, x_{n-1}]$ is true in M if and only if for every $a_0, \ldots, a_{n-1}$, the sentence

$$\varphi_{x_0, \ldots, x_{n-1}}[a_0, \ldots, a_{n-1}]$$

obtained from φ be substituting $a_0, \ldots, a_{n-1}$ for each free occurrence of $x_0, \ldots, x_{n-1}$, respectively, is true in M. We write

$$M \models \varphi$$

if φ is true in M.

We call M a *model* of a theory T with language L if every non-logical axiom φ is true in M. If M is a model of T, we write

$$M \models T.$$

A formula φ is called *valid* in a first-order theory T if for every model M of T, $M \models \varphi$. If φ is valid in T, we write

$$T \models \varphi.$$

Mathematicians call a sentence φ a theorem of T if φ is true in all models of T. This is neither a finitary method nor a mechanical procedure. It is not finitary because one has to verify φ in all models of T. Also note that any traditional proof of a theorem is finite where in an abstract model of T, using some logical deductions, φ is verified in M. It is quite natural to think that any proof of φ would have used only finitely many axioms of T and only finitely many logical rules of inference which are purely mechanical in nature.

This is indeed so. In a remarkable contribution to logic, Gödel defined a purely mechanical notion of proof and showed that a formula is true in all models of T if and only if it has a proof in T. We now present Gödel's definition of a proof in T and state his famous completeness theorem.

Fix a first-order language L. The following are *logical axioms* of L:

(a) **Propositional axioms**: These are formulas of the form $\neg A \vee A$.
(b) **Identity axioms**: These are formulas of the form $x = x$, where x is a variable.
(c) **Equality axioms**: These are formulas of the form

$$y_1 = z_1 \to \cdots \to y_n = z_n \to fy_1 \ldots y_n = fz_1 \ldots z_n$$

or formulas of the form

$$y_1 = z_1 \to \cdots \to y_n = z_n \to py_1 \ldots y_n \to pz_1 \ldots z_n.$$

(d) **Substitution axioms**: These are formulas of the form $A_x[t] \to \exists x A$, where A is a formula and t a term substitutable for x in A.

Rules of inference of L are

(a) **Expansion Rule.** Infer $B \vee A$ from A.
(b) **Contraction Rule.** Infer A from $A \vee A$.
(c) **Associative Rule.** Infer $(A \vee B) \vee C$ from $A \vee (B \vee C)$.
(d) **Cut Rule.** Infer $B \vee C$ from $A \vee B$ and $\neg A \vee C$.
(e) **∃-introduction rule**: If x is not free in B, infer $\exists x A \to B$ from $A \to B$.

Note that each logical axiom is true in every structure of L. Further, if the hypothesis of a rule of inference is true in a structure M of L, the formula inferred from them is also true in M.

Let T be a first-order theory with language L. A *proof* in T is a finite sequence of formulas $\varphi_1, \ldots, \varphi_n$ such that for every $1 \leq i \leq n$, φ_i is a logical axiom of L or a non-logical axiom of T or can be inferred from $\{\varphi_j : j < i\}$ by a rule of inference. In this case, we call $\varphi_1, \ldots, \varphi_n$ a proof of φ_n and φ_n a *theorem* of T. If φ is a theorem

6 Preliminary Concepts and Terminologies from Logic

of T, we write $T \vdash \varphi$. A theory T is called *consistent* if no sentence of the form $\neg \varphi \wedge \varphi$ is a theorem of T. If a theory T is not consistent, it is called *inconsistent*.

A sentence φ is said to be *undecidable* in T or *independent* in T if neither φ nor $\neg \varphi$ is a theorem of T. If the sentence T is not undecidable in T, we call the sentence *decidable* in T. A theory T is called *complete* if it is consistent and every sentence of T is decidable in T.

The following are two equivalent forms of Gödel's completeness theorem for first-order logic.

Theorem 6.0.1 (Gödel's completeness theorem, First Form) *Let φ be a formula of a first-order theory T. Then*
$$T \models \varphi \leftrightarrow T \vdash \varphi.$$

Theorem 6.0.2 (Gödel's completeness theorem, Second Form) *Let T be a first-order theory. Then T is consistent if and only if T has a model.*

We close this section by defining a class and a class model. Essentially a *class* is a formula $\varphi[x, w_0, \ldots, w_{n-1}]$ of ZF together with sets $z_0, \ldots, z_{n-1}$. We think of it as a collection of all sets x such that $ZF \vdash \varphi[x, z_0, \ldots, z_{n-1}]$, i.e., in informal notation
$$Z = \{x : ZF \vdash \varphi[x, z_0, \ldots, z_{n-1}]\}.$$

In this case, we say that the formula $\varphi[x]$ with parameters $z_0, \ldots, z_{n-1}$ defines the class Z. We write $a \in Z$ if
$$ZF \vdash \varphi[a, z_0, \ldots, z_{n-1}].$$

For instance, the formula $x = x$ stands for the class of all sets which is generally denoted by V. A class need not be a set. It is shown earlier that V and ON are not sets. On the other hand, every set z is a class—defined by the formula $x \in z$.

For each formula ψ of ZF, we define its *relativization* ψ^φ to the class $\varphi[x] = \varphi[x, z_0, \ldots, z_{n-1}]$ by induction on the length of ψ as follows:
(a) If ψ is atomic, then $\psi^\varphi = \psi$.
(b) If $\psi = \neg \eta$, then $\psi^\varphi = \neg \eta^\varphi$.
(c) If $\psi = \eta \vee \xi$, then $\psi^\varphi = \eta^\varphi \vee \xi^\varphi$.
(d) If $\psi = \exists x \eta$, then $\psi^\varphi = \exists x (\varphi[x] \wedge \eta^\varphi)$.

If Z is a class defined by a formula $\varphi[x]$ with parameters $z_0, \ldots, z_{n-1}$, often we write ψ^Z in place of ψ^φ. If X is a set defined by the formula $\varphi[x] = x \in X$ with parameter X, then also we write ψ^X instead of ψ^φ.

Now let X be a set and $\psi[x]$ a formula with parameters in X. The set
$$\{a \in X : ZF \vdash \psi^X[a]\}$$
is called a *definable subset* of X. Consider
$$\mathcal{D}(X) = \{A \subset X : A \text{ is a definable subset of } X\}.$$

$\mathcal{D}(X)$ is called the *definable power set* of X. Clearly, $\{z\} \in \mathcal{D}(X)$, $z \in X$, and $X \in \mathcal{D}(X)$. Also, it is easy to check that if X is finite, $\mathcal{D}(X) = \mathcal{P}(X)$. We know that $|\mathcal{P}(\mathbb{N})| = 2^{\aleph_0}$. However, the set of all formulas of ZF is countable. Therefore, since $\mathbb{N}$ is countable, $|\mathcal{D}(\mathbb{N})|$ is countable. Hence, there are subsets of $\mathbb{N}$ which are not definable.

Exercise 6.0.3 Let X be an infinite set. Show that $|\mathcal{D}(X)| = |X|$.

Since $\mathcal{D}(X)$ is defined by a quantification over formulas, it is not clear that in ZF one can prove that for sets X, $\mathcal{D}(X)$ are sets. However, Gödel defined a finite set of set operations such that $\mathcal{D}(X)$ is the closure of X under those finitely many set operations. This proves that for every set X, $\mathcal{D}(X)$ is a set. See [2].

Let T be ZF or ZFC or $ZFC + CH$ or $ZFC + \neg CH$, etc. Let X be a class defined by a formula, say $\varphi[x]$ with parameters $z_0, \ldots, z_{n-1}$. Then X is called a *class model* of T if for every non-logical axiom ψ of T, $ZF \vdash \psi^{\varphi}$.

As probably one of the best ever results proved in mathematics, Gödel could define a sentence CON_{ZF} of ZF meaning the metastatement "ZF is consistent." He then proved.

Theorem 6.0.4 (Gödel's Second Incompleteness Theorem)

$$ZF \nvdash CON_{ZF}.$$

Thus, the consistency of ZF is not a theorem of ZF. In entire mathematics, it is assumed that ZF is consistent. In order to show that a theory T such as above is consistent, one starts with the class V of all sets and then defines a class X such that every axiom ψ of T is true in X meaning that $ZF \vdash \psi^X$. For instance, assuming that ZF is consistent, Gödel defined a class L, called the class of all *constructible sets*, such that every axiom of $ZFC + GCH$ is true in L.

The class L of all constructible sets is not hard to define. By transfinite induction, we define $\alpha \to L_\alpha$, $\alpha \in ON$, as follows:

$$L_0 = 0 = \emptyset.$$

$$L_\alpha = \cup_{\beta < \alpha} L_\beta.$$

if α is a limit ordinal. Suppose $\alpha = \beta + 1$ is a successor ordinal. In this case, we define

$$L_\alpha = \mathcal{D}(L_\beta).$$

Since ON is not a set, $\alpha \to L_\alpha$ is not a function. However, there is a formula $\varphi[x, y]$ of set theory that defines the collection

$$\{(\alpha, L_\alpha) : \alpha \in ON\}.$$

The collection L of all constructible sets is the union $\cup_\alpha L_\alpha$. Hence,

$$L = \{x : \exists \alpha \in ON (x \in L_\alpha)\}.$$

Now it is easy to see that L is a class.

References

1. S.M. Srivastava, *A Course on Mathematical Logic*, 2nd edn. (Universitext, Springer, 2013)
2. K. Kunen, *Set Theory: An Introduction to Independence Proofs* (North-Holland Publishing Company, 1980)

Chapter 7
Zermelo–Fraenkel Set Theory

In this chapter, we define *Zermelo–Fraenkel set theory*, designated by ZF. This is a first-order theory with one binary predicate (relation) symbol $\in$, to be read as *belongs to*, only. Its axioms are listed below. To convey the content of the axioms better, we shall state the axioms informally in words also.

1. **Set Existence.** *There exists a set.*
   This is expressed by the following statement of ZF.
   $$\exists x(x = x).$$

2. **Extensionality.** *Two sets are the same if they contain the same sets:*
   $$\forall x \forall y (\forall z (z \in x \leftrightarrow z \in y) \rightarrow x = y).$$

   This axiom has a serious consequence: *Elements of a set are themselves sets*. For instance, the collection of all Indians and that of all Americans are not sets. If they were, by extensionality, they would be the same set because they contain no sets.

3. **Comprehension (subset) schema.** For each formula $\varphi[x, w_1, \ldots, w_n]$, the following is an axiom.
   $$\forall z \forall w_1 \cdots \forall w_n \exists y \forall x (x \in y \leftrightarrow x \in z \wedge \varphi).$$

   This axiom says that *given any "property of sets" expressed by a formula $\varphi[x, w_1, \ldots, w_n]$, for any fixed parameters $w_1, \ldots, w_n$ and any set z, there is a set y that consists precisely of those $x \in z$ that satisfy $\varphi[x, w_1, \ldots, w_n]$.*
   By extensionality, it can be proved that such a set y is unique, usually denoted by
   $$y = \{x \in z : \varphi[x, w_1, \ldots, w_n]\}.$$

   It is assumed that the variables x, y, z, and the w_i's are distinct.

4. **Replacement schema.** For every formula $\varphi[x, y, z, u_1, \ldots, u_n]$, the following formula is an axiom:

$$\forall z \forall u_1 \cdots \forall u_n (\forall x \in z \exists! y \varphi \to \exists v \forall x (x \in z \to \exists y (y \in v \wedge \varphi))),$$

where $\exists! y \varphi$ abbreviates the formula

$$\varphi \wedge \forall u (\varphi_y[u] \to u = y).$$

This axiom together with comprehension says that the range of a "function" on a set z that is defined by a formula φ is a set.

5. **Pairing.** *Given sets x and y, there is a set z that contains both x and y:*

$$\forall x \forall y \exists z (x \in z \wedge y \in z).$$

This axiom together with comprehension helps us to talk of sets of the form $\{x\}, \{x, y\}, \{x, y, z\}$, etc.

6. **Union.** *Given any set x, there is a set y that contains all those z that belong to a member of x:*

$$\forall x \exists y \forall z \forall u (u \in x \wedge z \in u \to z \in y).$$

This axiom together with comprehension will imply that the union of a family (i.e., a set) of sets is a set.

7. **Power Set.** *Given any set x, there is a set y that contains all subsets z of x:*

$$\forall x \exists y \forall z (\forall u (u \in z \to u \in x) \to z \in y).$$

This axiom together with comprehension will enable us to define the power set of a set.

8. **Infinity.** Based on the axioms introduced so far, in Chapter 1 we showed that the empty set exists, which we denote by $\emptyset$. The following formula is called the axiom of infinity:

$$\exists x (\emptyset \in x \wedge \forall y (y \in x \to y \cup \{y\} \in x)).$$

Without the infinity axiom, we can't prove the existence of an "infinite" set. Set $s(y) = y \cup \{y\}$. Take a set x that satisfies the infinity axiom. By the subset axiom, there is a unique set ω such that

$$\forall z (z \in \omega \to (z = \emptyset \vee \exists y \in \omega (z = s(y)) \wedge s(z) \in \omega)).$$

By taking $\mathbb{N} = \omega$, $0 = \emptyset$ and $s|\mathbb{N}$ as the successor function, using axioms of ZF we can prove that all Peano axioms are satisfied.

9. **Foundation.** This is the most unintuitive axiom. It is the following formula:

$$\forall x (\exists y (y \in x) \to \exists y (y \in x \land \neg \exists z (z \in x \land z \in y))).$$

It says that *the binary relation $\in$ is well founded on every non-empty set*. Using this axiom it is quite easy to prove that there is no sequence of sets $\{x_n\}$ such that for all n, $x_{n+1} \in x_n$. Now, if possible suppose there is a set x such that $x \in x$. Then take $x_n = x$, $n \in \mathbb{N}$, to arrive at a contradiction.

Since we use the axiom of choice whenever necessary, we state the axiom of choice formally below.

Axiom of Choice, AC.

$$\forall x (\forall y (y \in x \to y \neq \emptyset) \to$$

$$\exists f (\text{func}(f) \land \text{domain}(f) = x \land \forall y \in x (f(y) \in y))),$$

where $\text{func}(f)$ is the formula

$$\forall v \forall w_1 \forall w_2 ((\{v, \{v, w_1\}\} \in f \land \{v, \{v, w_2\}\} \in f) \to w_1 = w_2\}),$$

and

$$\text{domain}(f) = \{v : \exists w (\{v, \{v, w\}\} \in f)\}.$$

For any $v \in \text{domain}(f)$, the unique w such that $\{\{v, \{v, w\}\} \in f\}$ is denoted by $f(v)$.

Zermelo–Fraenkel set theory with choice is obtained by adding AC as a new axiom to ZF. It is denoted by ZFC.

Index

Symbols
$\mathbb{N}$, 4
$<_{\text{lex}}$, 31
$\emptyset$, 3
$\mathcal{D}(X)$, 126
$\mathcal{P}(X)$, 13
σ-field, 95
σ-ideal, 27
$\exists$-introduction rule, 124
$A \triangle B$, 20
A^*, 21
AC, 14, 131
B^A, 10
CON_{ZF}, 126
DC, 15
GCH, 58
L, 126
$M \models \varphi$, 123
$M \models T$, 123
ON, 52
$T \vdash \varphi$, 125
$T \models \varphi$, 124
V^*, 87
WOP, 45
$X^{<\mathbb{N}}$, 30
X/G, 22
ZFC, 131
ZF, 129
ZL, 23

A
Accumulation point, 37
Aleph, 57
Algebraically independent, 92
Algebraic closure, 91
Algebraic number, 68
Algebraic over a field, 92
Analytic sets, 103
Antisymmetric relation, 20
Associative rule, 124
Asymmetric relation, 20
Atomic formula, 122
Atomic σ-field, 97
Atom of a σ-field, 97
Axiom of choice, 14, 131

B
Banach–Tarski paradox, 119
Basis of a vector space, 85
Beth cardinals, 59
Bijection, 9
Binary relation, 19
Bound occurrence, 122

C
c.c.c., 32
Cantor-Bendixson derivatives, 55
Cantor ternary set, 99
Cardinality of a set, 56
Cardinal number, 55
Cardinal number of a set, 56
Chain in a poset, 23
Characteristic function, 13
Class, 4, 125
Class model of a theory, 126
Closed formula, 122
Closure of a formula, 123
Coanalytic sets, 103

Cofinality, 62
Co-finite set, 25
Co-meager set, 27
Co-null set, 27
Compact space, 77
Complete linear order, 32
Complete measure, 98
Complete theory, 125
Composition of functions, 9
Comprehension axiom, 129
Consistent theory, 125
Constructible sets, 58, 126
Continuous probability measure, 108
Continuum, 58
Contraction rule, 124
Countable, 12
Countable axiom of choice, 15
Countable chain condition, 32
Countable ordinal, 50
Countably generate field of sets, 95
CUB, 63
Cut rule, 124

D
Decidable sentence, 125
Definable power set, 126
Definable subset, 125
Definition by induction, 43
Definition by transfinite induction, 44
Dependent choice, 15
Derivative of a set, 37
Derived set, 37
Dimension of a vector space, 86
Distribution function, 107
Divisible group, 89
Domain of a binary relation, 19
Domain of a function, 9
Dual of a vector space, 87

E
Empty set, 3
Equality axiom, 124
Equivalence class, 20
Equivalence relation, 20
Even ordinal, 50, 54
Expansion rule, 124
Extensionality axiom, 2, 129

F
f.i.p., 25
Field of a binary relation, 20

Field of sets, 94
Filter, 25
Filter base, 25
Finite intersection property, 25
Finite ordinal, 50
Finite set, 16
First-order language, 121
First-order theory, 122
Forcing, 58
Formula, 122
Foundation axiom, 6, 52, 131
Free occurrence, 122
Free ultrafilter, 26
Function, 6

G
Generalized continuum hypothesis, 58
Group action, 22

H
Hahn–Banach theorem, 88
Hahn Decomposition Theorem, 105
Hamel basis, 87
Hausdorff's theorem, 112

I
Ideal in a ring, 89
Ideal on a set, 27
Identity axiom, 124
Inconsistent theory, 125
Independent sentence, 125
Indicator function, 13
Inductive set, 3
Infinite sets, 16
Infinity axiom, 3, 130
Initial ordinal, 55
Initial segment, 41
Injection, 9
Interpretation, 123
Invariant function, 21
Invariant set, 21
Inverse of a function, 9
Irreflexive relation, 20
Isolated point, 32, 37

L
Least upper bound axiom, 32
Lebesgue measure, 98
Lebesgue σ-field, 98
Lexicographic order, 31
Limit cardinal, 56

Limit ordinal, 50, 54
Linear functional, 87
Linearly independent set, 85
Linearly ordered set, 31
Linear map, 86
Linear order, 31
Linear span, 85
Locally finite, 79
Logical axioms, 124
Loset, 31
Lub axiom, 32

M

Maximal element, 23
Maximal filter, 25
Maximal ideal, 89
Meager set, 27
Measure, 98
Minimal element, 23
Model of a theory, 123

N

Natural numbers, 4
Negative set, 104
Non-logical axioms, 122
Nowhere dense, 27

O

Odd ordinal, 50, 54
Open formula, 122
Open set in a loset, 32
Orbit, 22
Orbit space, 22
Order dense, 32
Ordered pair, 5
Order isomorphic, 32
Order isomorphism, 32
Order topology, 82
Order type, 53
Ordinal number, 49, 53

P

Pairing axiom, 130
Paracompact spaces, 81
Partially ordered set, 22
Partial order, 22
Permutation, 12
Permutation group, 12
Point finite, 79
Poset, 22

Positive set, 104
Power set, 13
Power set axiom, 130
Principal ultrafilter, 26
Product of a family of sets, 14
Product topology, 78
Proof by induction, 42
Proof by transfinite induction, 42
Proof in a theory, 124
Proper ideal, 27
Propositional axiom, 124

Q

Quantifier-free formula, 122
Quotient space with respect to an equivalence relation, 20

R

Radon–Nikodym Theorem, 106
Range of a binary relation, 19
Range of a function, 9
Reflexive relation, 20
Regular cardinal, 63
Regularity axiom, 6
Relativization, 125
Replacement axiom, 130
Rules of inference, 124
Russell paradox, 5

S

Sentence, 122
Separable linear order, 32
Set existence axiom, 3, 129
Signed measure, 103
Singular cardinal, 63
Standard model of Peano arithmetic, 6
Strict linear order, 31
Strict partial order, 22
Structure for a first-order language, 122
Subformula, 122
Subset, 3
Substitutability, 122
Substitution axiom, 124
Successor cardinal, 56
Successor ordinal, 50, 54
Superset, 3
Surjection, 9
Suslin line, 35
Symmetric difference, 20
Symmetric relation, 20

T
Terms of a language, 122
Theorem of a theory, 124
Theory, 122
Topological support, 108
Torsion element, 89
Torsion-free group, 89
Total relation, 20
Trace measure, 104
Trace σ-field, 104
Transcendence basis, 93
Transcendence degree, 93
Transcendental number, 68
Transcendental over a field, 92
Transitive relation, 20
Transitive set, 51
Tree, 18
Trichotomy law, 31
Tychonoff plank, 83

U
Ultrafilter, 25

Ultraproduct of a family of sets, 27
Ultraproduct of fields, 90
Uncountable, 12
Uncountable ordinal, 50
Undecidable sentence, 125
Union axiom, 130
Universe, 1

V
Valid, 124

W
Well orderable, 39
Well ordered set, 39
Well ordering principle, 45

Z
Zermelo–Fraenkel set theory, 129
Zorn's Lemma, 23

SPRINGER NATURE

GPSR Compliance

The European Union's (EU) General Product Safety Regulation (GPSR) is a set of rules that requires consumer products to be safe and our obligations to ensure this.

If you have any concerns about our products, you can contact us on ProductSafety@springernature.com

In case Publisher is established outside the EU, the EU authorized representative is:

Springer Nature Customer Service Center GmbH
Europaplatz 3
69115 Heidelberg, Germany

The manufacturer's authorised representative in the EU is Springer Nature Customer Service Centre GmbH, Europaplatz 3, 69115 Heidelberg, Germany. If you have any concerns regarding our products, please contact ProductSafety@springernature.com

Printed and bound by CPI Group (UK) Ltd, Croydon, CR0 4YY

26/03/2026

02078916-0011